W9-ANF-484

# The Science Times Book of

# ARCHAEOLOGY

Other books in the series

*The Science Times Book of Birds*
*The Science Times Book of the Brain*
*The Science Times Book of Fish*
*The Science Times Book of Fossils and Evolution*
*The Science Times Book of Genetics*
*The Science Times Book of Insects*
*The Science Times Book of Mammals*

# The Science Times Book of ARCHAEOLOGY

EDITED BY
NICHOLAS WADE

THE LYONS PRESS

The Lyons Press, 123 West 18 Street, New York, New York 10011.

10 9 8 7 6 5 4 3 2 1

Library of Congress Cataloging-in-Publication Data

The Science times book of archaeology / edited by Nicholas Wade.
p. cm.
ISBN 1-55821-893-9
1. Archaeology 2. Antiquities. 3. Excavations (Archaeology).
I. Wade, Nicholas. II. Science times.
CC165.S36 1999
930.1—dc21 98-52875
CIP

# The Science Times Book of ARCHAEOLOGY

# Contents

# Introduction

Archaeology's realm lies beneath the surface. Its practitioners retrieve the raw material of their investigations from the grave or sepulchre or shipwreck, and sometimes from middens and trash heaps. Occasionally they are rewarded, like Heinrich Schliemann rummaging in the ruins of Troy or Howard Carter breaking into the untouched tomb of Tutankhamen, with some breathtaking treasure that escaped the notice of contemporary plunderers. But usually their pickings are bones and shards, blackened timber and charred stone, stuff so pitiable and broken that no one wants it.

The authors of historical written documents cannot be heard, but you can imagine them talking, perhaps in the next room. The makers of archaeological artifacts are not so close or familiar. Even their language is often unknown. They are dim, shadowy figures, summoned for a moment from their netherworlds by the archaeologist's skill to give fleeting glimpses of long-vanished civilizations.

Though its discoveries can almost make the dead come alive—at least in imagination—archaeology is no black art. It is a special kind of history, informed by scientific techniques of increasing power and subtlety. Far more information can now be extracted from a site than used to be the case, and methods of preservation now stabilize artifacts that once would have crumbled into dust.

With new methods of analysis and new search techniques, archaeologists continue to make finds even in the well-excavated regions of Mesopotamia, Mesoamerica, Egypt and classical Greece and Rome. But wholly new discoveries are not at an end. In the last few years there have been several spectacular archaeological events.

Chief among them is the finding in the Ardeche Valley region of France of a new cave whose walls are covered with prehistoric paintings at least 30,000 years old. Excavation of the cave has hardly begun, but it promises to open a new and unexpected window onto the culture of human societies a mere 20,000 years after the emergence of behaviorally modern humans and 25,000 years before the beginning of recorded history.

Another finding, even more dramatic in its way, was the recovery of Iceman, the body that emerged in 1991 from a thawing Alpine glacier. Judged to be some 5,300 years old, the perfectly preserved body bore tools, clothes and other perishable organic objects that seldom survive in normal archaeological sites. These forgotten technologies lend a sharp insight into the way of life of human societies at the threshold of civilization.

The past can still astonish and make news. The articles in this book first appeared in the Science Times section of the *New York Times,* many of them by our chief archaeology reporter, John Noble Wilford. With my colleagues on the science desk, I thank Lilly Golden of the Lyons Press for the idea for this book and for allowing these pieces a second life between hard covers.

# 1

# BEFORE HISTORY BEGINS

Skeletons indistinguishable from those of modern humans first appear in the fossil record from about 100,000 years ago. But stone tools and art objects start to accompany the skeletons only from about 50,000 years ago. Some archaeologists draw a distinction between the anatomically modern humans and the behaviorally modern humans, suggesting that some profound change, maybe genetically based, took place around this time, the finishing touch that made our forebears fully similar to ourselves.

If the distinction is valid, then prehistory can be thought of as starting about 50,000 years ago. On the scale by which paleontologists measure time, with life starting 3.5 billion years ago and the first animals appearing some 600 million years ago, 50,000 years is a few ticks of the watch's second hand. But in terms of human life, it represents some 2,500 generations.

The role of archaeology is to reconstruct the past, and nowhere is its task harder than in retrieving the story of the generations who lived before the era of written records. Almost every trace has completely vanished. Their bodies, buildings, clothes, every artifact made of organic material, have been subsumed in nature's remorseless recycling mechanisms. Almost all we have is a few bones and some stone tools on which to reconstruct nine tenths of human history.

Yet the picture has not been quite as completely erased as might seem at first glance. Geneticists have begun to extract the trove of information about early human migrations that lies embedded in our collective DNA. And two spectacular recent discoveries—the Ardeche Valley cave and the Iceman of the Tyrolean Alps—have opened quite unexpected new windows onto the darkest eras of prehistory.

Archaeology is not just about the old bones and dusty artifacts of museum exhibits. As the following articles suggest, it is also a field rich with novelty and surprise.

---

# Humans' Earliest Footprints Discovered

THE EARLIEST FOSSILIZED footprints of an anatomically modern human being have been discovered in 117,000-year-old sandstone on the shore of a South African lagoon, scientists reported.

Fossil bones may inspire paleontologists, and a particular type of DNA may satisfy geneticists that they have traced modern *Homo sapiens* back to a certain time in Africa. But nothing is more evocative of human ancestors than living, walking people leaving a trail of footprints.

The new discovery is particularly welcome in the study of human origins because the fossil record—in bones or footprints—is woefully incomplete for the period when archaic *Homo sapiens* evolved into the modern species. Most paleoanthropologists, and especially geneticists, think this fateful transition occurred between 100,000 and 200,000 years ago. Geneticists think the "African Eve," the one common ancestor of all living humans, lived at about this time.

Until now, only about 30 ancestral human fossils from that period—and no footprints—have been found anywhere in the world, mostly in southern Africa.

Dr. Lee Berger, who led the research team making the discovery, said, "These footprints are traces of the earliest of modern people."

Dr. Berger, a paleoanthropology professor at the University of Witwatersrand in Johannesburg, described the findings in an interview and at a National Geographic Society news conference in Washington. Articles on the research are published in *National Geographic* magazine and in the *South African Journal of Science.*

The most famous ancient footprints, made some 3.5 million years ago, were left by two adults and a child, presumably members of the *Australopithecus afarensis* species, who walked across a plain now known as Laetoli in Tanzania. The long track of prints was discovered in the late

1970s by an expedition led by Dr. Mary Leakey, the noted Kenyan fossil hunter. The prints provided further evidence that human ancestor species were walking upright long before they started making stone tools.

Earlier that year, another paleoanthropologist at the University of Witwatersrand, Dr. Ron Clarke, found ankle and foot bones of a 3.5-million-year-old prehuman ancestor, which are giving scientists their best evidence yet of the skeletal transition that facilitated upright walking. This individual, and presumably the Laetoli walkers, had a human-like ankle but an ape-like foot.

Other known prehuman footsteps were left 1.5 million years ago by *Homo erectus,* a predecessor species, at the Koobi Fora site in Kenya. Until now, no one had found any tracks of modern humans from the transition period before 100,000 years ago.

"We don't have any footprints in this time period," said Dr. Ian Tattersall, an authority on human origins at the American Museum of Natural History in New York City. "So it's interesting to find these and see that they are modern-looking, but not surprising."

Dr. David Roberts, a South African geologist, was the first to lay eyes on the three footprints in rock within 20 feet of the edge of Langebaan Lagoon, about 60 miles north of Cape Town. He was acting on a hunch. He had been picking up rock fragments that had apparently been chipped by human ancestors. And the surrounding gray sandstone was marked by animal tracks.

"I began searching for hominid footprints—and found them," he said. "Hundreds of people had walked over that area, including scientists, and not noticed the prints."

The human track extended only five feet, before disappearing at the base of a stone wall. Two of the prints are well preserved, revealing in detail the shape of the individual's big toe, ball, arch and heel. In every aspect, Dr. Berger said, these were the prints of modern human feet. The big toe was longer than the others, a trait known as the Egyptian toe; for many but not all people today, the big toe is the same length or slightly shorter than the next toe.

Several different dating techniques were applied in analyzing the rock bearing the prints, and they all agreed on an age of 117,000 years, give or take a few thousand years.

Judging by the length of stride and the foot size, eight and a half inches long, the individual was no more than five feet, six inches tall, probably shorter. Scientists said the person could have been a female or a juvenile or a small male.

Not much to go on, but enough for scientists to conjure up a gossamer image of the long-ago moment, soon after a rainstorm. An individual walked down the slope of a dune, approaching the lagoon at an angle and leaving tracks in the wet sand. Was this a casual stroll, or someone combing the beach for food washed up by the storm? Nothing in the steps betrays purpose or destination. But when the dune dried out, wind filled the footprints with sand, protecting them for eventual petrification and then discovery.

After finding the tracks, Dr. Roberts uncovered a number of stone tools that must have been made and used by the same people to kill and butcher prey and prepare skins. These tools included blades, scrapers, a projectile point and a large rock core from which flakes were struck. Other explorations in the region produced pieces of ocher pigment; the person who left the prints may have painted her or his body, an early example of self-expression.

Although some of the most spectacular recent fossil finds have been in Ethiopia and Kenya, South African scientists have been steadily gathering evidence suggesting that the modern human species emerged in their part of Africa and then spread north, eventually migrating to Asia and Europe.

Some of the strongest fossil-bone evidence of anatomically modern *Homo sapiens* has been collected at the mouth of the Klasies River, 375 miles from the footprint site. A cave there was occupied by early human hunters between 60,000 and 120,000 years ago. So the footprints, Dr. Berger said, "correspond very nicely with the Klasies people in time and location."

An African origin for modern humans is supported by recent studies of mitochondrial DNA, genetic material that is passed only through females. Measuring variations in mitochondrial DNA in different populations today, scientists have concluded that all humans are descended from one common female ancestor who lived in Africa between 100,000 and 200,000 years ago—the hypothethical "Eve."

"It's highly unlikely, of course, that the actual Eve made these prints," Dr. Berger said. "But they were made at the right time on the right continent to be hers."

—John Noble Wilford, August 1997

# Prehistoric Art Treasure Is Found in French Cave

IN THE MOUNTAINS OF SOUTHERN FRANCE, where human beings have produced art for thousands of years, explorers have discovered an underground cave full of Stone Age paintings, so beautifully made and well preserved that experts are calling it one of the archaeological finds of the century.

The enormous underground cavern, which was found on December 18, 1994, in a gorge near the town of Vallon-Pont-d'Arc in the Ardeche region, is studded with more than 300 vivid images of animals and human hands that experts believe were made some 20,000 years ago.

In this great parade of beasts appear woolly-haired rhinos, bears, mammoths, oxen and other images from the end of the Paleolithic era, creatures large and small and variously drawn in yellow ochre, charcoal and hematite.

The murals have surprised specialists because they also include a rare image of a red, slouching hyena and the era's first-ever recorded paintings of a panther and several owls. Specialists say this ancient art gallery surpasses in size that of the famous caves of Lascaux, also in southern France, and Altamira, Spain, which are widely held to be western Europe's finest collection of Stone Age art.

Archaeologists said they were thrilled not only by the number and the quality of the images but also by the discovery that the great underground site, sealed by fallen debris, appears to have been left undisturbed for thousands of years. They see this as tantamount to finding a time capsule full of hidden treasures.

One remarkable find, they said, was the skull of a bear, placed on a large rock set in the middle of one gallery against a backdrop of bear paintings.

"Is this some kind of altar? Someone placed the skull there for a reason," said Jean Clottes, France's leading rock art specialist. Many other skulls and bones of cave bears were found in the underground warren, along with bones, flint knives, footprints and remains of fireplaces, all of which archaeologists hope will provide important clues to the questions: What was the purpose of these paintings? What did their makers have in mind?

"Here we have a virgin site, completely intact. It may well change our perception, our thinking about the purpose and the use of cave art," said Mr. Clottes. At all other Stone Age sites found in Europe, he said, the floor and many objects had been disturbed by explorers.

A measure of the importance France attaches to the find is that the Minister of Culture, Jacques Toubon, himself chose to announce it at a news conference in Paris, in the company of France's top archaeologists.

"This discovery is of exceptional value because of its size and variety and because it was found undisturbed," said Mr. Toubon. "It is the only totally intact cave network from the Paleolithic era. It will help us to understand how human symbolism evolved."

He added that the site would not be open to the public in the foreseeable future but was now placed under government protection and accessible only to archaeologists. One expert said that until the climate inside the warren of caves has been stabilized, only three people may be inside at any one time.

Lascaux and other important prehistoric caves in France allow only a limited number of visitors, so as to reduce damage from moisture. Some sites are closed altogether and have nearby replicas for tourists.

The first inkling of a great unknown cave came when two men and a woman were exploring in the gorges of Ardeche, an area known for its decorated ancient caves and shelters.

"At one point we felt a draft coming out of the ground," said Jean-Marie Chauvet, a government guard of prehistoric sites, one of the three explorers. "For us that's a sign there is something else."

He said they took much of that day clearing fallen debris until they could enter a narrow hole that led to a greater space. They returned again the next week. Christian Hillaire, an amateur explorer, said the team first crawled through the narrow tunnel they had cleared, which was seven yards long. "Then we saw the first red markings on the walls with our helmet lights. So we kept going."

As the three lowered themselves on a rope, it turned out, they were entering through the ceiling of a great cavern. "There we began to see human markings and drawings everywhere," said Mr. Hillaire. "It was a great moment. We all shouted and yelled."

On their next visit, the team brought plastic to cover the ground where they walked to protect evidence. "We had seen footprints of bears," Mr. Hillaire said.

As the explorers advanced, they discovered a great cavern consisting of four main halls connected to one another by smaller galleries. The first hall they entered had only red paintings while in another hall all the murals were drawn in black.

Archaeologists who have visited the cavern since then and made the first videotapes, said that they have not yet fully explored the cavern for fear of disturbing the site. They said they believe there is more art as yet undiscovered.

The known part of the cavern consists of four great halls, up to 70 yards long and 40 yards wide, which are connected by smaller galleries roughly five by four yards, according a report issued by the Ministry of Culture. The more than 300 paintings and engravings vary in size between two feet and 12 feet long. Some stand alone, while others are clustered in panels or painted with some cohesion, such as two rhinos head to head, as if in a fight.

As Mr. Clottes showed a videotape revealing a panel with four horses' heads close together, drawn in charcoal, he said: "These are one of the great marvels of prehistoric art." The artist or artists, he continued, made use of the natural colors of the rock and of natural stone relief to give form and bulges to the animals.

But Mr. Clottes said there were many unanswered questions. Why were some paintings superimposed on older ones? Why was there almost an entirely different set of animals compared to other caves? At the Lascaux cave, which had a similar fauna at the time, he said, the images were overwhelmingly of bison and horses, but here bears and rhinos—animals that man did not usually hunt or eat—predominated.

Also, why were images of an owl, a panther and a hyena found in this cave but not in others in the Mediterranean region?

Mr. Clottes, who has visited the site, said that bears had evidently entered the cavern after the paintings were made because some paintings had clearly been scratched by bear claws. Humans and animals entering in the past most likely had used the natural entrances still visible in far corners, although they are now blocked by debris and clay.

Wanda Diebolt, a director of archaeology at the Ministry of Culture, described other puzzling markings. There were drawings that looked like an abacus: clusters of thick red dots, two inches or more in diameter. "We have found them in other grottoes, but we don't understand their symbolism," she said.

Archaeologists have long believed that the deep prehistoric caves were used not as habitats, because they were too dark, but for religious services or cults about which next to nothing is known. Some specialists believe that initiation rites were held in the caves and that the drawings of the human hands from the Paleolithic found here and elsewhere were a token of membership in a cult or community.

The Ministry of Culture is still waiting for radiochemical dating of the newly found art gallery. But several French archaeologists said they had no doubt about its authenticity and had estimated its age at 20,000 years on the basis of other, well-studied Paleolithic art of a similar style. They said that the images had the color, texture and hardened surface of ancient art.

The Ministry of Culture, in a report, cited other recent accomplishments for French prehistoric research. Among the new findings, it said, are important new datings for the so-called Cosquer cave, a richly painted cave near Marseilles, the entrance of which is underwater. It was discovered by a diver in 1991. While the Cosquer paintings at first were believed to date back 18,000 years, the latest carbon tests have established that the earliest paintings found there are some 27,000 years old.

Explorations of Cosquer in 1994 have also revealed that the total number of animal figures is more 125, while at least 55 stencils of human hands have been found, twice as many as known before.

—MARLISE SIMONS, January 1995

## Newly Found Cave Paintings in France Are the Oldest, Scientists Estimate

Scientific tests have shown some of the masterly drawn beasts discovered last December in the Chauvet—so named for Jean-Marie Chauvet, a government guard for prehistoric sites and a member of the exploration team—in the southeastern Ardeche to be at least 30,000 years old, making them the world's oldest known paintings.

The Culture Ministry said French and British specialists had determined that charcoal pigments of two rhinoceroses and a bison were between 30,340 and 32,410 years old.

The Culture Ministry said the test results have "overturned the accepted notions about the first appearance of art and its development," and show that "the human race early on was capable of making veritable works of art." Until now, experts have generally thought that early drawing and painting began with crude and clumsy lines and became more sophisticated only over centuries.

The Chauvet results were obtained through 12 radiocarbon datings from eight samples. They were carried out by two French institutes, the Center for Low Radioactivity at Gif-sur-Yvette and the Center for Radiocarbon Dating of the University of Lyons, and at Oxford, England, at the Research Laboratory for Archaeology and Art History.

Scientists have penetrated deeper into the cavern to search for new art and other signs of ancient human life and have found a fifth chamber with paintings.

The explorers have also discovered new creatures in ochre, hues of charcoal and red hematite. Researchers have now documented and photographed close to 300 animals, and say there may be more.

The work has been painstakingly slow, for workers can advance only on hard or rocky soil. They must sidestep the many soft and spongy areas in the humid cavern in order not to disturb vital evidence.

The most alluring new find, according to Mr. Clottes, the leader of the exploration, is a black drawing of a composite creature with the head and the hump of a bison standing upright on human legs. The archaeologists call it the Sorcerer.

"It's an extraordinary figure," Mr. Clottes went on. "Such composites are very rare in Paleolithic art. There may be less than half a dozen examples."

Next to the Sorcerer's left knee is an unusual curved triangle shaded in black, which Mr. Clottes said "has the appearance of a woman's vulva." He said two other such triangles were discovered, "images that we find mysterious."

Other new and surprising pictures include that of a young mammoth with feet that look like snowshoes made of large shaded circles, and several rhinos with broad black bands around their middles.

Mr. Clottes, who is 60, says he cannot get enough of the magnificent panels full of paintings. On five occasions, he has squeezed himself through the narrow entrance tunnel into the underground warren.

He has spent a particularly long time studying one of the most intriguing finds: a stone slab with the skull of a bear placed on it, as though it were an altar. Two other skulls are lying right at the foot of it, with another 20 nearby, he said.

"Were children playing there?" he asked. "Or is this related to a ceremony involving bears? These will be among the points we would like to clarify."

As the inventory of the great art gallery expands, archaeologists at the Culture Ministry say they are astonished that more than half the images are of dangerous animals like lions, rhinos and mammoths, rather than the horses and bison that early man hunted.

"This difference is important," Mr. Clottes said. "It shows that the beliefs, the attitudes toward animals of earlier people have changed over time. Perhaps this points to an evolution of their myths."

Specialists in France's rich rock art have marveled at what they call the sophistication of the techniques the artists used to present motion and perspective.

Some animals are interacting or fighting or stepping on each other. The head of a bison, for example, is drawn on the curve of a rock and is turned to obtain a double effect of perspective. Shading is used to give shape to the figures. Some figures are staggered, one behind the other, to obtain greater perspective.

Mr. Clottes said he would like another 20 radiocarbon datings. "One cannot just scrape a sample of a painting and damage it." He had lifted samples off the two fighting rhinos and a great bison where he found charcoal in a crack or in a lump. Other samples, he said, were taken from the ground in different sections of the cave and more from torchmarks on the walls.

The samples from the soil proved to be newer, dated at between 23,000 and 29,000 years ago. Charcoal taken from torchmarks dated from about 26,000 years ago.

"We are still finding many remarkable things," Mr. Clottes said, explaining that one of the torchmarks was made over the calcite that had naturally covered a painting made long before.

"This suggests that someone came into the cave some four thousand years after the painter and made that torchmark," he said. "But those early dates are the most intriguing. We have so few direct datings."

—MARLISE SIMONS, June 1995

# Vast Stone Age Art Gallery Is Found, But Dam May Flood It

ALONG THE BANKS OF THE COA RIVER in northern Portugal, in a spot where only shepherds' paths tell of the presence of humans, a prehistoric art gallery has been discovered on the rock face, setting off excitement, nervousness and accusations of a cover-up in the world of archaeology.

The images form an alluring parade of more than 60 animals, like bison, horses, ibexes and deer, and archaeologists estimate that they were chiseled into the rock face with sharp stone tools 20,000 years ago.

Historians have described it as the most important site of outdoor art of the Stone Age in Europe, revealing human settlements in inland areas previously thought too inhospitable for the coast-hugging early inhabitants of the Iberian Peninsula. But there, some 80 miles from the Atlantic coast, animal images were found, grouped in clusters, stretching for two miles along a deep gorge. Historians see this open-air site as important evidence that Stone Age art was not mainly made indoors, or in what are held to be the human race's most ancient sanctuaries: the caves or rock shelters where most other mural art from the Paleolithic era has been found.

Although the surprise of discovering a huge open-air gallery of Paleolithic art inevitably raises questions about its authenticity, the dozen or so experts who have seen the engravings so far say they have no reason to suspect forgery. "I believe it's genuine and probably one of the most important, if not the most important, outdoor Paleolithic site we know of," said Jean Clottes, France's leading rock art specialist, who was called in by the Portuguese government.

Yet, coinciding with the discovery has come the knowledge that the rock carvings may soon be lost. One portion, the lowest tier, has already been flooded. It was engulfed by water from a hydroelectric dam in the River Douro 12 years ago. Now a second dam is being built to create a backup reservoir and, unless the project is halted or changed, the remaining images will disappear under 300 feet of water in four years.

Indeed, it is only by chance that the existence of the art near Vila Nova de Fozcoa has become publicly known. An archaeologist identified the first engravings more than two years ago, but the find was kept a secret by the state-owned electricity utility, Electricidade de Portugal, which apparently did not want to disrupt construction of the dam, which started in September. More surprisingly, the prehistoric murals were also kept secret by the government's archaeology institute, the Institute for Architectural and Archaeological Patrimony, whose president visited the site a year ago.

Even now the survival of the art has not been secured. And in recent weeks archaeologists have spent as much time campaigning to save the engravings as analyzing them.

"We want the dam project to be stopped and the Coa Valley turned into a major cultural resource," said Joa Zilhao, a professor of archaeology at Lisbon University. The valley, he said, also holds later engravings and paintings from 3000 to 1000 B.C. and remains from Roman and medieval settlements.

"Some of the images are exceptionally beautiful," he said in a telephone interview after visiting the Paleolithic site. "There are some unusual scenes, like two horses touching and some animals running. They're not in the more common frozen positions."

He believes that the Stone Age people who made them probably moved up the valley of the River Douro, of which the Coa is a tributary, and probably lived in a harsh climate in an open, steppe-like landscape with many wild animals that provided food. "This was the peak of the Ice Age," he said. "These highlands could not have been very comfortable."

The discovery and its apparent cover-up for more than two years has now become a scandal in Lisbon, the capital, with historians and other intellectuals this month publishing open letters and demanding a parliamentary inquiry.

But around Hell's Canyon, as the gorge is known, the farmers and shepherds apparently are surprised by the fuss about scratchings in the rocks that they had long known about. "People of the region knew about the pictures," said Mila Simoes, a rock art specialist who has visited the area three times in recent weeks. "They treated Hell's Canyon as a magic place. They said the old people always told them that those rocks were inhabited by spirits. That children should not go there or touch anything."

The ancient art gallery carved into outcroppings and walls is reachable only by river boat or by hiking along the steep and windy shepherds' paths. Yet evidence that local people knew of the markings, Ms. Simoes said, is offered by the contemporary carvings and scratchings in the rocks, some of them close to the prehistoric images. In one case, she said, a modern visitor damaged an ancient engraving by retracing its lines.

Archaeologists first learned of the discovery less than two months ago from Nelson Rebanda, a 33-year-old archaeologist from the area. With financing from the electricity utility, Mr. Rebanda, a government employee, had been quietly studying and photographing the images, retracing a number of them on paper for the last two years.

"Suddenly I got a call from Nelson Rebanda in the middle of the night," said Ms. Simoes, the rock art specialist, who lives in Lisbon. "It was November seventh. He sounded in a panic. He was shouting that I should come quickly, that there was rock art that would soon be inundated." Because Mr. Rebanda had sounded so worried, she said, she and her husband, Ludwig Jaffe, also an archaeologist, made the seven-hour trip the next day.

"What we saw was amazing, magnificent," said Ms. Simoes, who has worked elsewhere in Europe and South Africa. She said she had learned that Mr. Rebanda had called her because the Douro dam downstream was briefly lowering its water level and would bare many of the flooded engravings for a few days. Mr. Rebanda told her he wanted her as a witness to testify that the rock images were genuine. "He said he would publish a thesis and a book, but by then the images would be covered by water, so he needed to authenticate them."

When Ms. Simoes demanded that the discovery be announced "because the Portuguese people must decide if they want a dam there or not," she went on, "Rebanda made a scene." She said: "He started shouting. He even threw his hat on the ground and jumped on it. He said he wanted nobody to know, that he wanted to keep all this for his book."

Once Ms. Simoes alerted her colleagues and the local press, Mr. Rebanda and his employers at the government archaeology institute changed their account of the events several times. At first, they announced that "dozens of Stone Age animal designs" had been discovered in recent weeks. But in a telephone interview, Mr. Rebanda conceded that he had identified the first engravings in 1992 and that he had reported those and subsequent findings to the institute in 1993 and 1994. Although he had several meetings with officials of the archaeology institute and the electricity utility—"we even talked about moving some of the rocks somewhere else"—he said that neither body "seemed interested in what I had to say."

In November 1993, he said, he asked officials from the utility to lower the waters in the downstream Douro dam to enable him to study the flooded images. "They told me that was too expensive," he said. He said he also suggested other solutions, like creating a dry dock around the rocks or, failing that, making underwater explorations.

Asked why his own institute did not back him, Mr. Rebanda said, "The people responsible there are architects. They are not much interested in archaeology."

Nuno Santos Pinheiro, the president of the institute, said he could not discuss the dates of the discoveries by telephone because this was "private, internal information." He said he visited the site last January. Asked why he did not announce the discoveries then, he said, "It would have been premature because the archaeologist had not found most of them. There were just a few." Second, he said, "we did not want to disclose this for security reasons, in order to protect the site."

The utility has said it did not know how to evaluate the art but had proceeded anyhow and started construction of the dam on schedule in September. Antonio Ribeiro dos Santos, a spokesman, said, "We are experts in electricity, not in archaeology." The archaeology report, he added, pointed out that the ancient images "were valuable" but it "never underlined that they were of global significance."

Clearly embarrassed, the Portuguese government declared the prehistoric site of the Coa Valley a national monument on December 13, which gives the Minister of Culture the right to halt the dam project. But construction on the $300 million reservoir continues and already employs 2,000 workers.

A citizens' movement is now forming to block the project. It has found some influential allies, among them a former undersecretary for energy, Nuno

Ribeiro da Silva, who during his tenure refused to approve the dam and has called the reservoir a "useless toy for engineers." The wine growers of the Oporto Valley also oppose it, fearing it will change the area's microclimate.

"We don't know yet if this is a drama of corruption or of ignorance," said Professor Zilhao. "But we must find out what happened."

While a group of archaeologists visited the site this month, no one is certain yet of the number of images. Mr. Rebanda, who has spent the most time there, says that he has counted close to 60 figures and that as many as 30 or more may have remained hidden underwater, even when he saw the river level at its lowest. He said he had made tracings of some 20 animals, some of them already submerged at the richest site at Hell's Canyon. "I did not stop anyone from taking pictures or filming," he said. "But I requested that I be the first to publish the material, which is normal."

Mr. Clottes, the rock art expert from France, has estimated that the engravings date back 20,000 years. After visiting the site on December 16, he said by telephone: "We have no direct dating, but to me this is undoubtedly from the Upper Paleolithic." Several other archaeologists who visited the site agreed that the images were ancient. "You can tell a lot from the patina," said Mr. Clottes, who has explored many prehistoric caves in France and is chairman of the International Commission of Rock Art. "In all the images I saw, the engraved lines are exactly as the rock itself, the same color, the same texture. Except in two, where people have redrawn some lines. You can see the difference.

"We also go by the style," Mr. Clottes said. "It matches the style of work that we have been able to date directly. There is that absence of details. Just one leg per pair. The hooves are never depicted. The horns of the aurochs are frontal while the head is in profile. And there are no human figures and none of the geometric designs you see at a later date."

The images are hammered or scratched into the rock and mostly range from six inches to three feet wide. Some are as much as six feet wide, like a long-maned horse and several aurochs, or bison, that became extinct in Europe in the seventeenth century.

"This gives us a whole different understanding of the art of that period," Mr. Clottes said. "It means that Paleolithic art is no longer cave art. It was best preserved in caves. But here we have a very big outdoor site. It's the biggest example we have."

Archaeologists have called for urgent test excavations in the area to further date it through other signs of human presence, like tools, bones or charcoal. Only two other sites with art from the Stone or Ice Age have been found in Portugal: in a cave at Escoural in the south and a small patch at Mazouco, in the Vila Nova region, which has just a few open-air animals. Interestingly enough, it was Mr. Rebanda who as a student first reported the Mazouco site, near his home village, in 1981, but it was his professor who got the credit. Colleagues of Mr. Rebanda have said that this time he was determined to keep the credit for himself.

"Whatever happens, the engravings must be preserved and not be damaged," Mr. Clottes said of Vila Nova. "There is no easy solution. If the dam is stopped, it means the images will be exposed to the public. They will be difficult to protect, in a remote place, scattered over a wide distance. And if they build the dam, it should be emptied every ten years to see what is happening."

—Marlise Simons, December 1994

# "Venus" Figurines from Ice Age Rediscovered in an Antique Shop

WHILE STROLLING ON THE RUE Notre Dame in Montreal, a young sculptor happened to look in the window of an antique shop and see a display of tiny statues carved in ivory and stone. It was the beginning of a rediscovery, revealed last week by archaeologists, that could lead to a better understanding of a distinctive but enigmatic form of Ice Age art.

The statuettes in the window, of nude women with exaggerated breasts and buttocks, were prized specimens of the first so-called "Venus" figurines to be excavated in the 1880s in caves near Monaco. Other discoveries across Eurasia followed, opening the eyes of scholars to a shared artistic expression that seemed to unite far-flung prehistoric people from the Pyrenees to Siberia. Perhaps these were symbols in fertility rites, keys to prehistoric sexual roles, expressions of a common mythology, or simply Ice Age pornography.

But several of these first figurines, from the Grimaldi site across the Italian border from Monaco, had been missing for nearly all this century.

The sculptor bought the figurines four years ago, drawn by their aesthetic allure and unaware of their significance. Last October, having decided to have them evaluated by experts, he put the five figurines—each about one inch to six inches tall—in a cigar box and took them to the Redpath Museum at McGill University in Montreal.

"I nearly fainted," recalled Dr. Michael Bisson, the McGill anthropologist who examined the contents of the cigar box. "It was obvious that these were Upper Paleolithic female figurines, and they were genuine. I knew we had a big find."

His first impression was confirmed a few days later by Dr. Randall White, an associate professor of anthropology at New York University who is a specialist in Ice Age art. The figurines were carved 18,000 to 25,000 years ago, in the late Stone Age. From written descriptions after their original discovery and other evidence, Dr. White concluded that these were indeed five of the seven missing figurines from the Grimaldi site.

Through the antique dealer, the other two missing pieces were soon recovered as well. Dr. Bisson said the previous owners of the artifacts did not want to be identified, and the name of the sculptor was being withheld for security reasons until all seven figurines could be transferred safely to the Mus e des Antiquites Nationales outside Paris.

On a recent visit to Montreal, curators of the French museum corroborated the initial identification of the artifacts and agreed to undisclosed terms for their acquisition, pending official approval this month. There the recovered seven pieces will be reunited with seven other Grimaldi figurines. The only other known specimen is at the Peabody Museum at Harvard University.

"These specimens constitute one of the two most significant discoveries of Paleolithic art in the past twenty-five years," Dr. Bisson said. The other, he said, was the discovery of art in an underwater cave off the southeast coast of France, the Grotte Cosquer.

Dr. White said the rediscovery could "completely transform our understanding of the total collection of Ice Age figurines." No more than 200 of them have been found at prehistoric sites separated by thousands of miles. Some differ in style from place to place, but apparently not in concept. Most of the known Venus figurines are from central Europe and Russia.

But it was at the Grimaldi site on a cliff facing the Mediterranean Sea that Louis Jullien, a French chemist and amateur archaeologist, made the first discoveries in 1883. For some reason, he kept the figurines a secret for at least a decade, finally selling seven of the 15 to the Paris museum. Some archaeologists at the time attacked the figurines as fakes. The other eight disappeared when Mr. Jullien retired and moved to Montreal, where he died in 1928.

The figurines and a collection of stone tools from the same site, Dr. Bisson said, remained in the Jullien family for many years. Harvard acquired its one figurine from a daughter. But the trail to the others seemed to grow cold, and scholars had lost interest in the quest.

Of the rediscovered pieces, Dr. White said, the most impressive is a double figurine made of pale green serpentine. It is only two inches tall and shows a pregnant woman with a featureless face and pendulous breasts aligned back to back with an unidentified animal, possibly a wolverine.

Five of the recovered figurines are of pregnant women. A sixth is the upper part of a woman's body, and the seventh may be a man. All their faces are blank, their limbs partly missing and their sexual parts often exaggerated. Men were seldom represented in this art form; it was more usual to see them depicted with animals in hunting scenes painted on cave walls.

The figurines were made of soapstone, serpentine and ivory, possibly from the tusks of woolly mammoths. Since these animals did not live this far south, the ivory indicated that people living near the Mediterranean must have had some contact with the mammoth hunters of northern Europe.

Because the specimens were never cleaned for museum display, Dr. White said, the cracks and knife marks still contain residue of sediment and charcoal. Flecks of charcoal are to be analyzed to obtain radiocarbon dates. By comparing the sediments with those from various levels in the caves, scientists should also be able to fix their dates fairly precisely. In this way, they hope to determine if variations in style indicate differences in time or in the purpose for which the figurines were carved.

Some scholars eschew the term "Venus" figurines, contending that this implies an interpretation that is not necessarily correct. The word "Venus" suggests a goddess and beauty, they say, but it is possible that neither is involved in this art.

For Dr. Patricia Rice, an anthropologist at West Virginia University in Morgantown, however, they still are Venus figurines. "Changing names is never very successful," she said, "so we might as well stick to Venus and make sure we understand what the meanings really are."

Until the 1980s, the favored interpretation related the figurines to fertility. But most fertility rites are associated with agricultural societies, and these were hunter-gatherers living thousands of years before crop cultivation. Besides, in an analysis of 180 figurines over a wide geographic range,

Dr. Rice found that a surprisingly small percentage of them appeared to be pregnant.

Dr. Rice suggested other possible interpretations, each involving a recognition of the important roles of women in the Ice Age culture. In a hunter-gatherer society, the men, who hunted, often came home empty-handed, which meant that it fell to the women, who gathered, to provide much of the food. Since many of the figurines had holes, and so must have been worn as pendants, men may have carried them on the hunt as reminders of hearth and home. But there is no evidence, she said, that the figurines meant that women were worshipped as goddesses.

Could this be pornography? "That's always possible," Dr. Rice said. "But I say no. If it were, the figurines would have been much more realistic."

One crucial piece of evidence remains missing, and without it, no definitive interpretation may ever be possible. No one knows if the carvers of these buxom females were women or men.

—JOHN NOBLE WILFORD, February 1994

# Iceman's Stone Age Outfit Offers Clues to a Culture

STILL NO ONE KNOWS who he was or what he was doing high in the Tyrolean Alps that day some 5,300 years ago, the day he died. No one can be sure of the quirks of nature that somehow mummified the corpse, then entombed it in a glacier and preserved it and his possessions so long in a semblance of a life only lately departed.

Of one thing scientists are now certain about the naturally mummified Alpine Iceman, whom hikers discovered in September 1991 in the melting ice on the Austrian-Italian border at an elevation of 10,530 feet. In the first genetic analysis of the body, they determined that he was European born and bred, closely related to modern northern and Alpine Europeans.

Hardly surprising, of course, for a man of that time when people seldom traveled far, but scientists said this finding should lay to rest lingering suspicions of a hoax. The possibility of an elaborate fraud, a Piltdown Man for the late Stone Age or Copper Age, has worried cautious scientists and prompted popular speculation. Could this be an Egyptian mummy planted in new surroundings? Or a pre-Columbian American mummy like those recently uncovered in the deserts of Peru and Chile?

An international research team, writing in the journal *Science,* said the genetic findings made "the possibility of fraud highly unlikely."

With the authenticity of the incredible Iceman thus more assured, scientists could feel more comfortable interpreting everything else about the discovery, especially the man's tools and weapons and the clothes he was wearing. These are the things archaeologists concentrate on in trying to reconstruct the lives of long-ago people, and usually all they have to go on are grave goods, possessions selected for accompanying the dead and

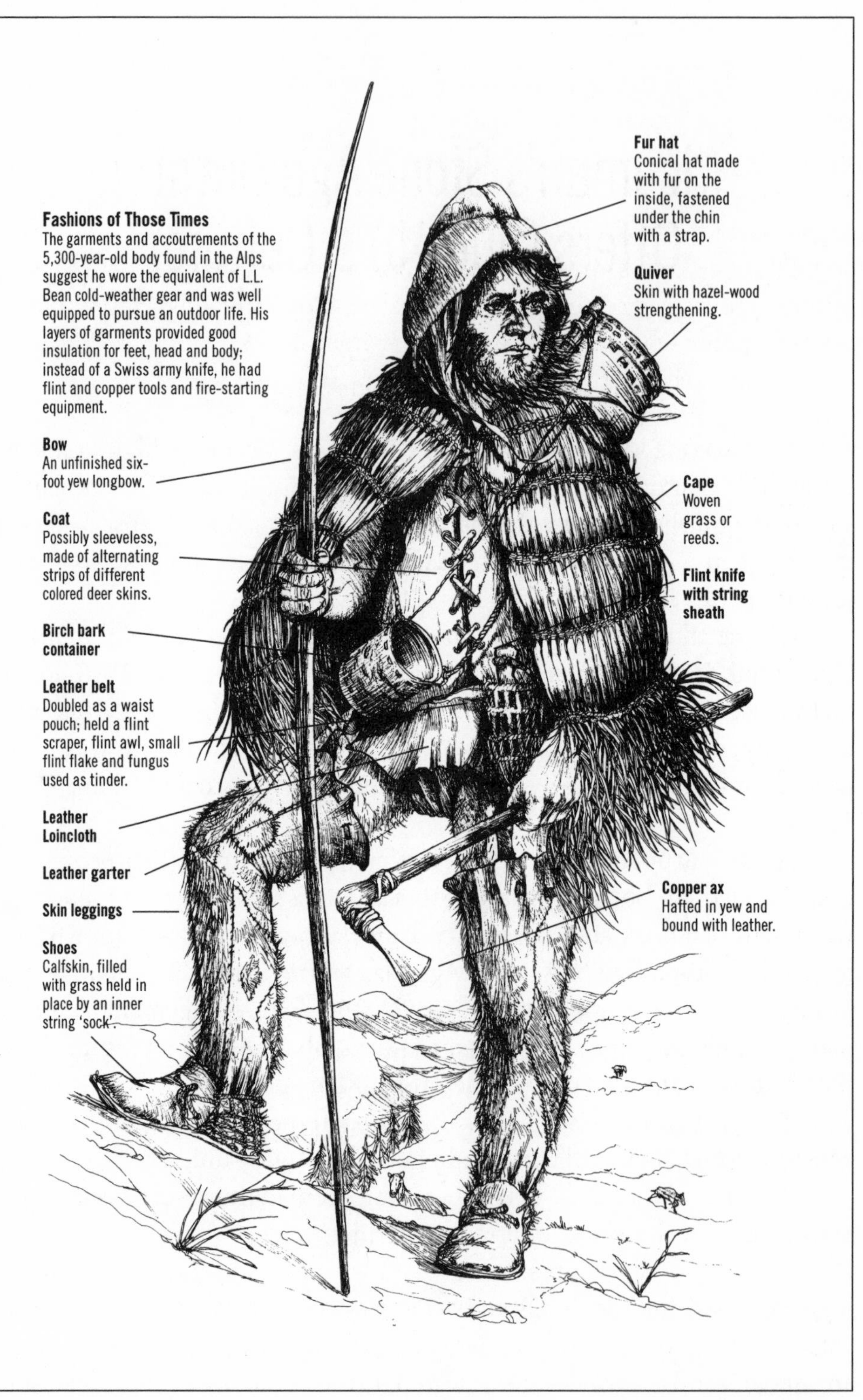


Michael Rothman

perhaps not representative of daily life. But here is a rare opportunity for looking at the practical equipment and clothing of one man in use at one moment of his life—and death.

Among the results of their research is a descriptive inventory of Alpine fashions in those remote times. Scientists may not be able to account for the man's presence on the mountain crest—was he a farmer, hunter, trader, prospector, village outcast or, more probably, a shepherd?—but they know what he was wearing, down to his underwear and garters.

Much of the reconstruction of his apparel from the seven preserved articles of clothing has been conducted by Dr. Markus Egg, an archaeologist at the Roman-Germanic Central Museum in Mainz, Germany. The results were reported in detail by Dr. Konrad Spindler in *The Man in the Ice,* translated into English and published in London by Weidenfeld & Nicholson. Dr. Spindler, an archaeologist at the University of Innsbruck in Austria, is directing the international team of 147 scientists investigating the Iceman. A summary and assessment of the clothing studies was included in a comprehensive review of all the research published in the British journal *Antiquity* by Dr. Lawrence Barfield, an archaeologist at the University of Birmingham in England who specializes in Alpine prehistory.

The Iceman was probably in his late twenties or thirties and was five feet, two inches tall, and in one respect, he would have been right in step with modern styles. He wore a leather waist pouch, not unlike today's popular "fanny packs." His foundation garment was a leather belt that included this pouch, into which he had stuffed a sharpened flint scraper or knife, a flint awl, a small flint flake and a dark mass of organic material probably intended for use as tinder in fire making.

The belt held up a leather loincloth, and leggings made of animal skin had been attached to it by suspended leather strips serving as garters. For his upper torso he had a jacket, possibly sleeveless, made from alternating strips of different-colored deer skin. Completing his ensemble was an outer cape of woven grasses or reeds of a type that, Dr. Barfield said, was still used in the Alps up to the beginning of this century. A conical cap, made with the fur on the inside, was originally fastened below his chin with a strap. His feet were protected from the cold by much-repaired shoes of calfskin filled with grasses for insulation.

Close examination of the apparel revealed the varied skills of this apparently self-sufficient man, particularly his expertise in treating animal skins and tanning them as leather. Researchers have identified in the clothing the skins of goat, calf, red deer and brown bear; roe deer, ibex and chamois may also be present.

Although much of the Iceman's equipment was described soon after the discovery, the list of 20 different items is now more definitive. "It is a contemporary mountain survival kit and more," Dr. Barfield wrote.

With the Iceman was an unfinished six-foot longbow made of yew. Why he would be on such a journey without a serviceable bow is one of the many puzzles. A quiver made of animal skin contained 14 broken or otherwise unserviceable arrows of viburnum and dogwood, two with flint tips and some with feather fletching. Other contents of the quiver included two sinews, perhaps Achilles tendons of a large animal, that probably were for the bowstring; a line made from tree fiber; a bundle of bone points wrapped in a leather thong; and a curved antler point, perhaps a needle.

More of his belongings included a frame made of hazel and skins, presumably a rucksack, and two sewn birch-bark containers that, from the blackened interior and maple leaves, may have been used for carrying embers for fire at the next campsite. There were also more flint tools and knives, a fragment of string net possibly for capturing birds and two pieces of birch fungus threaded onto a leather thong. This could have been a folk antibiotic, a kind of prehistoric penicillin, Dr. Barfield observed.

The Iceman also had with him a copper ax with a yew haft and leather binding. At first, the ax was thought to be bronze, which would have indicated the man lived somewhat more recently, perhaps only 4,000 years ago. Radiocarbon dating of plant remains and other material finally put the date at between 5,100 and 5,300 years ago. At this time, European culture remained essentially Stone Age with only the beginning of metal use, while in the valley of the Tigris and Euphrates rivers the early cities of Sumer were rising, wheeled vehicles were coming into use, wine was being fermented and civilization generally was taking shape.

Though a member of a somewhat more primitive culture, the Iceman and his cultural baggage have succeeded in impressing archaeologists, even surprising them.

"Because of ingrained ideas of progress, the sophistication of past technology and culture tends still to take us by surprise," Dr. Barfield wrote. "In reality, few of us today have any of the skills which most people would have had during the fourth millennium."

In his book, Dr. Spindler examined evidence about the man's body and proposed some tentative interpretations. Since X rays revealed some broken ribs, perhaps he got into a fight in the village below and this led to the injury and a hasty flight into the mountains. Dr. Barfield thought this unlikely. "A more economical explanation," he said, "would be that he was already up in the mountains with his sheep when an accident prevented his descent to the valley before the snow came." Or it could be that the ribs were broken in the rough handling when the frozen corpse was discovered and hauled down from the mountain.

The Iceman's teeth were heavily worn, the examination showed, which Dr. Spindler attributed to chewing dried meat. Or it could have been from the working of leather. It was not a complete surprise to find tattoos over the body, the practice going back well into antiquity. But Dr. Spindler noted that the simplicity of the tattoo designs and their inconspicuous placement mostly on the back and legs suggested that they may not have been decorative but may have been related to medical treatment akin to acupuncture.

Earlier, it was widely reported that the Iceman had been emasculated, leading to lurid speculation. This Dr. Spindler firmly quashed. "With the most careful intervention with sterilized rubber gloves," he reported, "it was possible millimeter by millimeter to lift the penis from the attached scrotum."

Even before the new genetic studies of the body, archaeologists pointed to one artifact that they felt certain refuted the idea of fraud. Among the Iceman's equipment was a small antler point inserted into a lime-wood handle. This was recognized as an instrument for retouching flint tools. But no one thinking to perpetrate a fraud would have thought to include this particular tool in the Iceman's kit; it was not known as an artifact type before this discovery, Dr. Barfield said, but since then, it has been identified among materials at several other sites of that time and region.

The analysis of DNA from samples of the mummy's muscle, bone and connective tissue was conducted independently by researchers from the

University of Munich in Germany and Oxford University in England, with contributions by scientists at the University of Zurich in Switzerland and the University of Innsbruck in Austria. The principal authors of the report in *Science* were Dr. Oliva Handt and Dr. Svante Paabo of Munich.

The researchers isolated genetic material passed down from generation to generation on the maternal side, known as mitochondrial DNA, and compared their patterns to similar material from representative populations of modern humans. The Iceman's DNA did not match any of the specimens from sub-Saharan Africans, Siberians or American Indians. The genetic patterns of the Iceman, the scientists concluded, "fits into the genetic variation of contemporary Europeans" and was "most closely related to mitochondrial types determined from central and northern European populations."

As Dr. Bryan Sykes, an Oxford geneticist on the team, said, "There are relatives of the Iceman all over northern Europe."

—John Noble Wilford, June 1994

# World of Ancient Iceman Comes into Focus

IT WAS IN 1991 THAT A vacationing German couple happened across a well-preserved body lying facedown in a slushy pocket of glacier in the Italian Alps. Since this chance discovery of what turned out to be a 5,300-year-old corpse, the world's oldest known human-flesh remains, scientists and archaeologists have teased out remarkable information on the man and his environment from the equipment and clothing found with him.

A bit of ember he carried to restart his campfire was from a tree likely to have grown south of the spot where he died. That evidence suggested he was on his way from the fertile Venosta Valley in northern Italy and had probably made his home there. A grain of domesticated wheat clinging to his fur clothing indicated that he had had contact with civilization, which in those days and parts would have been a small farming village.

Studies of the Iceman, as the body is known, have moved beyond his accoutrements to his flesh. His guardians at the University of Innsbruck in Austria, where he is being preserved at glacial temperature and humidity, have at last opened up the corpse itself to research. Using specially designed titanium instruments that leave no trace elements, doctors have snipped off tiny samples from the man's shrunken organs and tissues and delivered them to scientists in Europe and the United States for studies that are slowly building answers to the questions about the man's life and death.

In September 1995, Dr. Werner Platzer, the Innsbruck anatomist who oversees research on the body, announced that preliminary findings showed the man's stomach was empty when he died. But his large intestine contained considerable amounts of material.

"That means he had probably not eaten for eight hours," Dr. Platzer said in a telephone interview, adding that the contents of the large intestine were being investigated. The finding, which had been predicted by radiologists' interpretations of CAT scans, hinted that the man may have been hungry and weak when he died. A hypothesis that has strong support among scientists is that he died of hypothermia after being surprised by one of the sudden snowstorms that come up on the Hauslabjoch, the 10,500-foot pass over the main ridge where he was found.

Other new findings suggest that the man had not been in perfect health. Dr. Andrew Jones, an environmental archaeologist at the Archaeological Resource Centre in York, England, identified the eggs of a parasitic whipworm in a small sample from the man's colon. The preliminary results do not reveal how severe the infestation was, and Dr. Jones could not say whether the parasites caused the man any discomfort. Light cases often go unnoticed, but severe infestations carry severe symptoms, Dr. Jones said, and even a moderate infestation may have weakened the man.

Another potential vulnerability appeared in the man's lungs, where Dr. Raul J. Cano, a microbiologist at the California Polytechnic State University in San Luis Obispo, recently found a fungus called *Aspergillus,* most likely the species *fumigatus.*

"I know that *Aspergillus* has been associated with lung disease, but we have no reason to believe he died of anything other than natural causes," said Dr. Cano, who isolated the DNA of the fungus. And last year doctors found that the lungs were as black as a smoker's, probably a result of living in a shelter with an open hearth.

Discussion of what caused the man's death has been complicated from the start by X rays that show five broken ribs on his right side. Radiologists are unable to determine whether these fractures occurred before the man's death, under the weight of the glacier or during the rough recovery.

"There are so many rib fractures, they're angled, the chest is severely decreased in diameter and the posterior ribs are dislocated from the spine," said Dr. William A. Murphy, Jr., the head of diagnostic imaging at the University of Texas M. D. Anderson Cancer Center in Houston and one of the radiologists who has studied the thousands of X rays and CAT scans of the body. "It's my opinion that it would take significant force to do that, and I can imagine that force from the weight of ice."

But neither he nor Dr. Dieter zur Nedden, his Austrian colleague, believes the question will soon be laid to rest. That keeps alive the theory of

Dr. Konrad Spindler, the University of Innsbruck prehistorian who proposed that the man had been involved in a fight in his village then fled into the mountains, where he succumbed to his injuries.

There are, indeed, signs that the man's life had not been easy. Dr. Horst Seidler, an anthropologist at the University of Vienna, said the man may have lived through episodes of extreme hunger, illness or metal poisoning that arrested his growth. Examining X rays of the man's shinbone, Dr. Seidler and his colleagues found 17 Harris lines, thin layers of bony material that form in the hollow of a bone when growth stops. They calculated that severe disturbances occurred in the man's ninth, fifteenth, and sixteenth years.

"Possibly this had to do with periods of hunger in the transition between seasons," Dr. Seidler said.

One of Dr. Seidler's next projects is to compare tissue samples from the Iceman with those of the 500-year-old Peruvian girl discovered on an Andes mountain.

"These two finds were conserved in the same condition," said Dr. Seidler, who has seen the girl's remains in Peru. "This is the first time we've had material with which to compare him."

Other findings illuminate the man's activities during his life. Dr. Don Brothwell, an archaeologist at the University of York in England, announced that his team had detected unusually large amounts of copper on the surface of the man's hair. In a telephone interview Dr. Brothwell suggested that the prehistoric man, who was found with a copper ax, may have been involved in the processing of copper, the first metal smelted.

"What I'm hoping to do is check out if there is copper on the fingers and in the lungs," Dr. Brothwell said, noting that if the man were smelting he would have been breathing in copper particles.

In the Copper Age, a period that may have begun as early as 3,000 B.C. in the Alpine region, people searched for malachite, a copper carbonate that appears naturally as a bluish green efflorescence on rock and cliff sides. They scraped and flaked off the malachite and then smelted it in a crucible in a campfire. They increased the fire's temperature by blowing oxygen into it through bellows. The nearly pure copper would then be poured into a stone mold, for an ax, for example. Some prehistorians have wondered whether the man was in the mountains looking for malachite, though he was not found with digging tools.

"We can only speculate about how metal production was organized," said Dr. Lawrence Barfield, an archaeologist at the University of Birmingham in England who studies the Copper Age in northern Italy. "We don't know whether there were specialists doing it or whether it was something that anybody could do."

But even if the man did not make his copper ax, he still had to polish and sharpen it, and that process, too, could have left dust on his hands that he then rubbed into his hair, said Dr. Markus Egg at the Roman-Germanic Central Museum in Mainz, Germany, where the equipment and clothing have been conserved and studied.

While much of the remaining research is in the court of doctors and scientists, archaeologists, too, are still intrigued with the find. Archaeologists in Italy and Austria spent three summers looking for more signs of prehistoric human activity in the high valleys below the Hauslabjoch, and excavations have turned up several sites where hunters who lived millennia before the Iceman stopped to camp overnight, built a fire and left a few flint tools and flakes behind.

At least three of these sites were found along a path over which local shepherds still guide their sheep to summer pastures, and the Iceman was also found on this route, said Dr. Annaluisa Pedrotti, an archaeologist at the University of Trient in Italy.

At that time, Dr. Pedrotti pointed out, a person would not climb to such altitudes without sound economic or, possibly, religious reasons. Like most prehistorians involved in the project, both she and Dr. Egg think the man was probably a farmer or shepherd on his way to or from the summer pastures nearby. Yet even this interpretation has problems.

"We still don't have any typical shepherd equipment with him," Dr. Barfield said. "Where is his walking stick? Why isn't he wearing anything made of sheepskin?"

—BRENDA FOWLER, December 1995

# First Settlers Domesticated Pigs Before Crops

DIGGING AT THE RUINS of a village in southeastern Turkey, where people lived more than 10,000 years ago, archaeologists expected to turn up the usual traces of a society on the verge of the agriculture revolution. There should be leftover grains of wild wheat and barley and perhaps the bones of butchered sheep and goats in some early stage of domestication.

The archaeologists found nothing of the kind. Instead, to their complete surprise, they dug up the ample remains of pig bones.

The discovery, they said, strongly suggests that the pig was the earliest animal that people domesticated for food. The diminished size of the molars was one of several clues that the transformation of wild boars into pigs was under way at that time. Radiocarbon analysis put the date at 10,000 to 10,400 years ago.

So in the foothills of the Taurus Mountains at a site known as Hallan Cemi, the domestication of the pig appeared to have occurred 2,000 years earlier than once thought—and 1,000 years before the taming and herding of sheep and goats.

Much earlier, at least 12,000 years ago, wolves more or less invited domestication as the dog, developing a symbiotic relationship with people. They became camp followers, sentinels and "best friend." Only in a few cultures later on were dogs served as food.

A broader significance, archaeologists said, was the absence of any sign of wheat or barley at the settlement. The prevailing assumption, based mainly on research to the south in Syria and the Jordan River Valley, has been that with the end of the last ice age, wild grains were abundant in the more temperate climate over the entire Middle East. People settled down

## An Early History of Domestication

| Animal | Wild Progenitor | Principal Region of Origin | Approximate Date of Domestication |
|---|---|---|---|
| Dog | Wolf | Western Asia | 12,000 years ago |
| **Pig** | Boar | Turkey | 10,400 to 10,000 years ago |
| Goat | Bezoar goat | Western Asia | 9,000 years ago |
| Sheep | Asiatic mouton | Western Asia | 9,000 years ago |
| Cattle | Auroch | Western Asia | 8,000 years ago |
| Reindeer | Reindeer | Northern Eurasia | Unknown |
| Llama | Guanaco? | South America | 7,000 years ago |
| Alpaca | Llama species? | South America | 7,000 years ago? |
| Horse | Horse | Central Asia | 6,000 years ago |
| Donkey | Ass | Arabia, North Africa | 6,000 years ago |
| Water buffalo | Water buffalo | Southern Asia | 6,000 years ago |
| Dromedary camel | Camel | Arabia | 5,000 years ago |
| Bactrian camel | Camel | Central Asia | 5,000 years ago |
| Cat | Cat | Western Asia | 5,000 years ago? |
| Chicken | Jungle fowl | Southern Asia | 4,000 years ago |
| Guinea pig | Cavy | South America | 3,000 years ago? |
| Guinea fowl | Guinea fowl | North Africa | 2,300 years ago |
| Yak | Yak | Himalayas | Unknown |
| Mithan | Gaur | Southeast Asia | Unknown |
| Bali cattle | Banteng | Southeast Asia | Unknown |
| Rabbit | Rabbit | Iberia | Roman era |
| Goldfish | Carp | China | A.D. 960 |
| Turkey | Turkey | North America | A.D. 500 |

*Source: The Cambridge Encyclopedia of Human Evolution* (Cambridge University Press): Dr. Michael Rosenberg/University of Delaware.

to harvest them, and this led to agriculture, animal husbandry and eventually the rise of cities and civilization.

"All early agricultural models are predicated on the assumption that people gathered wild wheat and other grains," said Dr. Michael Rosenberg, an archaeologist at the University of Delaware and director of the Hallan Cemi excavations. "But this is the earliest settlement site so far north, and it has no cereals. So another resource must have made it possible to settle down."

In a report at a meeting of the Society for American Archaeologists, Dr. Richard W. Redding, a University of Michigan archaeologist and member of the discovery team, said that a heavy reliance on data from southern sites in the Levant might have resulted "in a very narrow view of the origin of food production." There may have been a variety of ways by which people made the transition from foraging to farming, and some of them did not include the intensive use of wild cereals as a crucial first step.

Dr. Patricia Wattenmaker, a University of Virginia archaeologist with wide experience in Middle East excavations, said the new findings were among the first from this part of the Turkish highlands in prehistory and were certain to force a serious rethinking of theories regarding human subsistence patterns leading up to agriculture.

"It looks as if the pattern varies from place to place," Dr. Wattenmaker said. "This takes the punch out of arguments that environmental factors" over a wider area triggered the transition toward agriculture, she added, and suggested that "cultural factors were really the key."

Dr. Robert J. Braidwood, a professor emeritus of archaeology at the University of Chicago who is a specialist in research on early agriculture, praised the Hallan Cemi excavations for providing a much-needed examination of preagricultural cultures beyond the Levant. But the Turkish village of no more than 150 inhabitants was extremely small, he cautioned, and evidence from three or four more sites in the area might be necessary before drawing any sweeping conclusions.

The Hallan Cemi site is scheduled to be flooded next year by a new dam on the Batman River.

In three years of excavations at Hallan Cemi, though, archaeologists established that people there had left the wandering life of hunting and gathering for a more sedentary village existence. Ruins of small stone houses and stone sculptures indicated a permanent settlement, and the growth pattern in freshwater clamshells at the site revealed year-round occupation. Evidence of long-distance trade in obsidian, copper and Mediterranean shells reflected the expansion of economic horizons by an increasingly complex society.

All this was happening at the time of the Natufians, people in the Jordan Valley who were probably the first to adopt settling down as a permanent way of life. But if wild cereals were critical to the Natufians' transition, the people at Hallan Cemi apparently depended on gathering nuts and seeds, hunting wild sheep and deer and raising pigs. The absence of any wild grains at the site was determined by Dr. Mark Nesbitt, a paleobotanist at University College, London.

No single piece of the pig evidence is conclusive, Dr. Redding reported, but all the clues together "are congruent with the early phases of the domestication of pigs."

Not only are the bones plentiful and the molars smaller, he said, but they show that the people appeared to favor young male pigs more than would be expected if they were hunting wild animals. A preponderance of the bones were of male pigs under one year of age. If they were raising pigs, they would spare most of the young females for breeding. Survivorship patterns of hunted animals reveal a more normal age distribution.

Pigs may have been the villagers' insurance against famine caused by any sudden shortage of nuts and fruits and wild game. In a preagricultural sedentary culture, Dr. Rosenberg said, such shortages posed a greater risk because the people had a more limited foraging and hunting range.

"We think they fiddled around with maintaining animals to decrease that risk," he said, "and pigs make sense if they are not gathering and growing grains."

For one thing, young pigs are easily obtained and tamed. They require little labor to control since they can be left to forage for themselves throughout the community. And they are the most efficient domesticated animal, Dr. Redding said, in that they convert 35 percent of food energy into meat, compared to 13 percent for sheep or a mere 6.5 percent for cattle.

Pigs, the archaeologist concluded, may have represented one more transitional step in some preagricultural societies; the pattern was not always a direct progression from settling down to growing cereals to raising animals. Perhaps the subsistence strategy of the highland villagers was to supplement their diets of nuts, fruits and grasses with pigs until cereal production was adopted. In time, Dr. Redding said, the highlanders took up grain cultivation, probably as an innovation borrowed from the south.

In any case, the archaeologists said, as soon as the people of Hallan Cemi began growing grain, there was a sharp decline in domestic pigs, which were gradually replaced by domestic sheep and goats. It was a necessity. Pigs compete with people for cereals. They could no longer be left to forage unattended near the village and fields, and they are not as easily herded as sheep and goats.

Although some of the interpretations are tentative and more research is required, Dr. Rosenberg and Dr. Redding said they were increasingly confident in their evidence for the early domestication of the pig. And contrary to previous findings in the Levant, they said, there could be sedentary village life without an abundance of grains, wild or cultivated.

"Hallan Cemi is almost a mirror image of what's going on at this time in the Levant," Dr. Redding said. "We will have to rethink all the models we've been developing about early food production."

—JOHN NOBLE WILFORD, May 1994

# Remaking the Wheel: Evolution of the Chariot

IN ANCIENT GRAVES on the steppes of Russia and Kazakhstan, archaeologists have uncovered skulls and bones of sacrificed horses and, perhaps most significantly, traces of spoked wheels. These appear to be the wheels of chariots, the earliest direct evidence for the existence of the two-wheeled high-performance vehicles that transformed the technology of transport and warfare.

The discovery sheds new light on the contributions to world history by the vigorous pastoral people who lived in the broad northern grasslands, dismissed as barbarians by their southern neighbors. From these burial customs, archaeologists surmise that this culture bore a remarkable resemblance to the people who a few hundred years later called themselves Aryans and would spread their power, religion and language, with everlasting consequence, into the region of present-day Afghanistan, Pakistan and northern India.

The discovery could also lead to some revision in the history of the wheel, the quintessential invention, and shake the confidence of scholars in their assumption that the chariot, like so many other cultural and mechanical innovations, had its origin among the more advanced urban societies of the ancient Middle East.

New analysis of material from the graves shows that these chariots were built more than 4,000 years ago, strengthening the case for their origin in the steppes rather than in the Middle East.

If the ages of the burial sites are correct, said Dr. David W. Anthony, who directed the dating research, chariots from the steppes were at least contemporary with and perhaps even earlier than the earliest Middle East chariots. The first hint of them in the Middle East is on clay seals, dated a century or two later. The seal impressions, from Anatolia, depict a light, two-wheeled vehicle pulled by two animals, carrying a single figure brandishing an ax or hammer.

"Scholarly caution tells me the matter is not resolved," said Dr. Anthony, an anthropologist at Hartwick College in Oneonta, New York. "But my gut feeling is, there's a good chance the chariot was invented first in the north."

While praising Dr. Anthony's work, Mary Littauer, an independent archaeologist and coauthor of *Wheeled Vehicles and Ridden Animals in the Ancient Near East* (Brill, 1979), was not ready to concede the point. "It's still debatable," she said. "A spoked wheel is not necessarily a chariot, only a light cart on the way to becoming chariots."

A chariot is usually defined as a lightweight vehicle with two spoked wheels and drawn by two horses. The earliest ones in the steppes and the Middle East probably used a form of the ox yoke around the horses' necks. Yokes adapted especially for horses, allowing them more freedom of movement, do not appear in Middle East art until the middle of the second millennium B.C. This probably represented the fully developed chariot, but the earlier versions also qualify as chariots in the eyes of nearly all scholars.

Other archaeologists and historians said they would not be surprised to learn that the chariot had originated in the steppes. After all, pastoralists there were probably the first to tame and ride horses; as Dr. Anthony determined in other research reported four years ago, this may have occurred at least 6,000 years ago. Then they developed wagons with solid disk wheels, and many centuries later learned to make the lighter spoked wheels, the breakthrough invention leading to the fast, maneuverable chariot.

The results of Dr. Anthony's dating research were described in interviews and in a report presented at a meeting of the American Anthropological Association. An interpretation of the results was published in *Archaeology,* the magazine of the Archaeological Institute of America.

The culture of the Russian and Kazakhstan steppes was virtually unknown until about 15 years ago, when Russian archaeologists began systematic excavations at several sites east of the Ural Mountains. One of the first sites explored was at a place called Sintashta, southeast of the city of Magnitogorsk. Another was Petrovka, 400 miles to the east on the Ishym River in northern Kazakhstan.

Archaeologists thus refer to these ancient people as the Sintashta-Petrovka culture. Their widely scattered settlements were linked culturally, as seen in the many similarities of their ceramics, metal weapons and tools, architecture and burial rituals.

At Sintashta, archaeologists uncovered a large settlement of about 50 rectangular structures arranged in a circle within a timber-reinforced earthen wall. They found slag deposits from copper metallurgy, bronze weapons and gold earrings and the remains of six chariots in a cemetery of elite graves covered by earth mounds, or kurgans. Similar artifacts, and more chariots, were discovered at several other sites.

Russian scientists estimated that the culture flourished between 1700 and 1500 B.C. This inspired Dr. Stuart Piggott, a retired archaeologist at Edinburgh University in Scotland and a specialist on ancient wheeled transport, to propose several years ago that if these people had developed chariots, as preliminary reports from the graves suggested, they could be the earliest developed anywhere.

But it was not until the end of the Cold War that Western scholars began to learn the details of these excavations and could test Dr. Piggott's hypothesis. Enter Dr. Anthony, who had been collaborating with Russian archaeologists on other research involving prehistoric cultures. He went to Dr. Nikolai B. Vinogradov, an archaeologist at Chelyabinsk State Pedagogical Institute, who directed excavations of the chariot burials, and got permission to apply new radiocarbon dating techniques on those materials.

In a grave pit at Krivoe Ozero, 120 miles north of Sintashta, archaeologists had found an assemblage of materials typical of these burials. The heads of two ritually killed horses were deposited next to the body of a man and a chariot, much like later Aryan practices. Other grave goods included spear points, a bronze ax and a dagger, three pots and four disk-shaped cheekpieces from a horse harness. The bit used to control a horse passes through these cheekpieces, which were made of bone and antler. The rein is attached to the bit.

For dating the burial, Dr. Anthony took four samples of bone from the skulls of two horses in a single grave. They were analyzed at the University of Arizona at Tucson by a highly accurate type of radiocarbon dating technology using an accelerator mass spectrometer. This yielded a range of age estimates, from 2136 to 1904 B.C., with an average of 2026—much earlier than the Russians had estimated. Similar ages were determined independently at Oxford University in England, using samples from other steppe sites.

The chariots in the graves had decayed to dust, but not without a trace. The wheels had been fitted into slots cut into the dirt floor of the burial chamber. The lower parts of each wheel left stains as they decayed. The stains preserved the shape and design of the wheels. Some parts of a chariot superstructure were also preserved in this way.

Sintashta-Petrovka wheels had eight to 12 spokes. Early chariots in the Middle East, as revealed in the Anatolian seal impressions, had only four spokes. The steppe chariots were also quite narrow. The distance between the two wheels was consistently less than four feet, probably suitable for only one person. Chariots widely used in Middle East warfare in later centuries were wider, capable of carrying two or three people.

Both the difference in numbers of spokes and in width, Dr. Anthony said, suggest that the steppe chariots probably evolved locally. He said it was improbable that the technology developed independently in both places; more likely, it arose in one place and was soon introduced in the other.

"It is likely that chariot designs and perhaps uses varied from region to region, even during the initial rapid diffusion of chariot technology," he said. For example, the steppe chariot might have been developed not for warfare, but for use in ritual races meant to settle disputes or win prizes, which was an Aryan practice.

Dr. James D. Muhly, a professor of ancient history at the University of Pennsylvania, said the findings were important because they revealed a moment in the transition from wagons and carts with solid wheels to lighter vehicles with the spoked wheel, which weighed a tenth as much as the solid versions. People in Mesopotamia and on the steppes had been hitching oxen, asses and horses to these heavy wagons and carts since before 3000 B.C. Mesopotamian art as early as 2600 shows warriors being transported to battle in carts with solid wheels.

With the light, two-wheeled chariot, said Dr. Robert Drews, a classics professor at Vanderbilt University in Nashville, people could fully exploit

| | |
|---|---|
| 4000 B.C. | The first evidence of man on horseback dates to this era. It was discovered in Ukraine, where pastoralists of the steppes were probably the first to tame and ride horses. |
| 3200 B.C. | The first wheels were solid wood disks. They were probably used on four-wheeled carts that could transport more goods than either a man or horse could carry. |
| 2000 B.C. | What may have been the first chariot wheels had eight to 12 spokes, a lighter and more efficient design. They were discovered in the steppes of Russia. |
| 1850 B.C. | Four-spoked chariot wheels depicted on seals found in Anatolia constitute early evidence of the use of chariots in the Middle East. |
| 1275 B.C. | The chariot wheel reached the height of its development. Chariots were used in the battle between the Egyptians under Ramses II and the Hittites. |

the horse as a draft animal. And while an ox cart traveled only two miles an hour, a team of chariot horses could cover 10.

This led to many applications, the most flamboyant of which was the chariot as a terrifying and efficient instrument of war in the late Bronze Age. From about 1700 to 1200 B.C., military strategy centered on chariotry, setting off an arms race. Rulers from Anatolia to Egypt and Crete to Mesopotamia strained palace treasuries to build more and more chariots, which were expensive. In the Bible, King Solomon is said to have paid 600 shekels of silver for each chariot; it had cost David only 50 shekels to buy a team of oxen and a threshing floor.

And then, as Dr. Drews pointed out in his book *The End of the Bronze Age,* published by Princeton University Press, there was the expense of maintaining the chariot force with horse trainers and grooms, veterinarians and carpenters, charioteers to drive and warriors skilled in archery and a bureaucracy of clerks and quartermasters to keep track of all this.

Although "the general character of chariot warfare remains unexplored," Dr. Drews wrote, the vehicle was probably used as a mobile shooting platform for archers. With charioteers at the reins, a mass of chariots would race forward while archers standing behind the driver would send a rain of arrows into the enemy ranks. It is thought that the Hittites introduced a third man, someone to hold a shield.

One of the culminating battles of chariotry came early in the thirteenth century B.C., when armies of the Hittites and Egyptians clashed on the plains of northern Syria at Kadesh. Muwatallis II, the Hittite king, deployed a force of 3,500 chariots, and Ramses II is supposed to have countered in kind, but the battle seems to have ended in a stalemate. By the end of that century, as armies learned to blunt the attacks with swarming infantry and later cavalry, the age of the chariot as a weapon drew to a close. The high-speed vehicle was reduced to roles in sport and regal parades.

Among the charioteers of the steppes, the pattern was much the same. Aryan-speaking charioteers, sweeping in from the north in about 1500 B.C., probably dealt the death blow to the ancient Indus Valley civilization. But a few centuries later, by the time the Aryans compiled the Rig Veda, their collection of hymns and religious texts, the chariot had been transformed to a vehicle of ancient gods and heroes.

Chariot technology, Dr. Muhly noted, seems to have left an imprint on Indo-European languages and could help solve the enduring puzzle of where they originated. All of the technical terms connected with wheels, spokes, chariots and horses are represented in the early Indo-European vocabulary, the common root of nearly all modern European languages as well as those of Iran and India.

In which case, Dr. Muhly said, chariotry may well have developed before the original Indo-European speakers scattered. And if chariotry came first in the steppes east of the Urals, that could be the long-sought homeland of Indo-European languages. Indeed, fast spoke-wheeled vehicles could have been used to begin the spread of their language not only to India but to Europe.

One reason Dr. Anthony has his "gut feeling" about the steppe origin of the chariot is that in this same period of widening mobility, harness cheekpieces like those from the Sintashta-Petrovka graves show up in archaeological digs as far away as southeast Europe, possibly before 2000 B.C. The chariots of the steppes were getting around, possibly before anything like them in the Middle East.

—John Noble Wilford, February 1994

# Enduring Mystery Solved as Tin Is Found in Turkey

ARCHAEOLOGISTS HAVE DISCOVERED tin in Turkey. No one is predicting a rush by miners to stake claims or any quick riches to be made on the world's metal markets, for the amount discovered is trifling. But scholars are hailing it as a solution to one of the most enduring mysteries about ancient technology: Where did the metalsmiths of the Middle East get the tin to produce the prized alloy that gave the Bronze Age its name?

The new findings could change established thinking about the role of trade and metallurgy in the economic and cultural expansion of the Middle East in the Bronze Age, which ran from about 3000 B.C. to 1100 B.C.

After thousands of years in which copper was the only metal in regular use, the rising civilizations of Mesopotamia set off a revolution in metallurgy when they learned to combine tin with copper—in proportions of about 5 to 10 percent tin and the rest copper—to produce bronze. Bronze was easier to cast in molds than copper and much harder, with the strength of some steel. Though expensive, bronze was eventually used in a wide variety of things, from axes and awls to hammers, sickles and weapons, like daggers and swords. The wealthy were entombed with figurines, bracelets and pendants of bronze.

Digging through ruins and deciphering ancient texts, scholars have shaped an image of the Bronze Age as a time of vibrant economic expansion, the earliest Sumerian cities and the first great Mesopotamian empires. They found many sources of copper ore and evidence of furnaces for copper smelting. But despite their searching, they could never find any sign of ancient tin mining or smelting anywhere closer than Afghanistan.

It seemed incredible that such an important industry could have been founded and sustained with long-distance trade alone. But where was there any tin closer to home?

After systematic explorations in the central Taurus Mountains of Turkey, an archaeologist at the Oriental Institute of the University of Chicago has found a tin mine and ancient mining village 60 miles north of the Mediterranean coastal city of Tarsus. This is the first clear evidence of a local tin industry in the Middle East, archaeologists said, and it dates to the early years of the Bronze Age. Some of the metal might have been imported from faraway Afghanistan or elsewhere, but not all.

In an announcement by the university, Dr. Aslihan Yener of the Oriental Institute reported that the mine and village demonstrated that tin mining was a well-developed industry in the region as long ago as 2870 B.C. She analyzed artifacts to re-create the process used to separate tin from ore at relatively low temperatures and in substantial quantities.

"Already we know that the industry had become just that—a fully developed industry with specialization of work," Dr. Yener said. "It had gone beyond the craft stages that characterize production done for local purposes only."

Dr. Vincent C. Pigott, a specialist in the archaeology of metallurgy at the University Museum of the University of Pennsylvania, said, "By all indications, she's got a tin mine. It's excellent archaeology and a major step forward in understanding ancient metal technology."

To Dr. Guillermo Algaze, an anthropologist at the University of California at San Diego and a scholar of Mesopotamian civilizations, the discovery is significant because it shows that bronze metallurgy, like agriculture and many other transforming human technologies, apparently developed independently in several places. Much of the innovation, moreover, seemed to come not from the urban centers of southern Mesopotamia, in today's Iraq, but from the northern hinterlands, like Anatolia, in what is now Turkey.

Speaking of the ancient tin workers of the Taurus Mountains, Dr. Algaze said, "It's very clear that these are not just rustic provincials sitting on resources. They had a high level of metallurgy technology, and they were exploiting tin for trade all around the Middle East."

The mine, at a site called Kestel, has narrow passages running more than a mile into the mountainside, with others still blocked and unexplored. The archaeologists found only low-grade tin ore, presumably the remains of richer deposits that had been mined out.

For this reason, Dr. James D. Muhly, a professor of ancient Middle Eastern history at the University of Pennsylvania, said he was skeptical of interpretations that Kestel was a tin mine.

"They have identified the geological presence of tin," he contended. "Almost every piece of granite has at least minute concentrations of tin in it. But was there enough there for mining? I don't think they have found a tin mine."

In her defense, Dr. Yener said, "His arguments are still based on an analysis of the mine and not the industry. He has to address the analysis of the crucibles."

On the hillside opposite the mine entrance, the archaeologists found ruins of the mining village of Goltepe. Judging by its size, Dr. Yener said, 500 to 1,000 people lived in the village at any one time. Radiocarbon dating of charcoal and the styles of pottery indicated that Goltepe was occupied more or less continuously between 3290 and 1840 B.C. It began as a rude village of pit houses dug into the soft sedimentary slopes and later developed into a more substantial walled community.

Scattered among the ruins were more than 50,000 stone tools and ceramic vessels, which ranged from the size of teacups and saucepans to the size of large cooking pots. The vessels were crucibles in which tin was smelted, Dr. Yener said, and they hold the most important clues to the meaning of her discovery and her answer to skeptics.

Slag left over from the smelting, collected from inside the crucibles and in surrounding debris, contained not low-grade tin ore but material with 30 percent tin content, good enough for the metal trade. This analysis, including various tests with electron microscopes and X rays, was conducted with the assistance of technicians from Cornwall, a region of England famous for tin mining since ancient times.

The tin-rich slag, Dr. Yener concluded, established beyond doubt that tin metal was being mined and smelted at Kestel and Goltepe. They could not have met all of the Middle East's tin needs in the Bronze Age, she said, but neither was all the tin imported, as had long been thought.

By this time, the scientists realized the significance of all the stone tools and could reconstruct the methods of those ancient tin processors.

The mining was done with stone tools and fire. Miners would light fires to soften the ore veins and make it easier to hack out chunks. Since the shafts were no more than two feet wide, the archaeologists said, children may have been used for much of the underground work. This inference was reinforced by the discovery of several skeletons buried inside the mine; their ages at death were 12 to 15 years. Further examination should determine if they died of mining-related illnesses or injuries.

Once extracted, the tin ore, or cassiterite, was apparently washed, much the way forty-niners in the American West panned for gold in streams, separating nuggets from the rest. Many of the stone tools at the site were used to grind the more promising pieces of ore into smaller fragments or powder.

Then crucibles, set in pits, were filled with alternating layers of hot charcoal and cassiterite powder. Instead of using bellows, workers blew air through reed pipes to increase the heat of the burning charcoal. Tests indicated that this technique could have produced temperatures of 950 degrees Celsius and perhaps as high as 1,100 degrees (1,740 to 2,000 degrees Fahrenheit), sufficient to separate the tin from surrounding ore.

Droplets of tin were encased in molten slag. When this cooled, workers again used stone tools to crush the slag to release the relatively pure tin globules. Sometimes the slag was heated again to separate any remaining tin.

If this site is typical of ancient tin processing, Dr. Yener concluded, then archaeologists may have overlooked other local sources of Bronze Age tin. They had been searching for the remains of large furnaces for tin smelting, much as had already been found for copper smelting, and had not suspected that a major tin-processing operation could be conducted successfully with fairly low grades of ore and in small batches in crucibles. In this manner, with hard work and many people, tin might even be recovered at relatively low temperatures.

The identity of these highland mining people is unknown, but their pottery betrays cultural ties to societies in northern Syria and Mesopotamia. The Taurus Mountains were known in the powerful cities of southern Mesopotamia as a rich source of metals, and Sargon the Great, founder of the Akkadian Empire in the late third millennium B.C., wrote of obtaining silver there. Although Kestel was close to many ancient silver, gold and copper mines, no traces of copper were detected at the site, indicating that the processed tin was traded elsewhere for the production of bronze.

Among the remaining mysteries of ancient metallurgy is the question of how people first recognized the qualities of bronze made from tin and copper and how they mixed the alloy.

For several centuries before the Bronze Age, metalsmiths in Mesopotamia were creating some tools and weapons out of a kind of naturally occurring bronze. The one used most frequently was a natural combination of arsenic and copper. The arsenic fumes during smelting must have poisoned many an ancient smith, and since the arsenic content of copper varied widely, the quality of the bronze also varied and must have caused manufacturing problems.

Scholars have yet to learn how the ancient Mesopotamians got the idea of mixing tin with copper to produce a much stronger bronze. But excavations have produced tin-bronze pins, axes and other artifacts from as early as 3000 B.C. In the Royal Cemetery at the ancient city of Ur, nine of 12 of the metal vessels recovered were made of tin-bronze, suggesting that this was the dominant alloy by the middle of the third millennium B.C.

Some of the answers may be awaiting discovery in Afghanistan. Sumerian texts referred to the tin trade from the east, and finally, in the 1970s, Russian and French geologists identified several ancient tin mines in Afghanistan. Until now, that discovery had seemed to resolve the issue of Mesopotamia's tin source. Since tin appeared to be more abundant there, archaeologists are eager to explore the mines because they may provide evidence of the first tin-bronze technologies, but the civil war in that area is keeping such plans on hold.

Although he was skeptical of Dr. Yener's claim to have found an ancient tin mine, Dr. Muhly praised her effort to find the sources of metals in the Middle East as "tremendously important archaeology" because of the connection between the development and widening use of bronze and the emergence of complex societies, large urban centers, international trade and empires.

The Bronze Age could not continue forever, scholars say, in part because tin was so hard to get, contributing to the expense of the metal alloy. The age came to an end around 1100 B.C., when iron, plentiful and accessible just about everywhere, became the most important metal in manufacturing.

—JOHN NOBLE WILFORD, January 1994

# Finding Suggests Weaving Preceded Settled Life

SOME 27,000 YEARS AGO, an innovative group of hunters and gatherers were in the habit of setting up their summer base camps near a river along the Pavlov Hills in what is now the southeastern Czech Republic. They mixed the fine soil with water and molded it into human and animal figurines and fired them, creating the oldest known fired ceramics. They took the two-and-a-half-million-year-old technology of flaking stone tools a step farther by grinding them into smoothly polished pendants and rings, the earliest known examples of ground stone technology in Europe.

At a meeting in Minneapolis of the Society for American Archaeologists, scientists announced that this same group, contemporaries of the earliest cave painters of France and northern Spain, has left the oldest evidence of weaving in the world. The site has yielded clay fragments bearing impressions of textiles or basketry, which according to Dr. James M. Adovasio of Mercyhurst College in Erie, Pennsylvania, and Dr. Olga Soffer of the University of Illinois at Urbana, push back the known origin of these technologies at least 7,000 years, to 27,000 years ago.

It also validates a suggestion long offered by some archaeologists that the origin of textile technology by far predates the Neolithic period of plant and animal domestication to which it had traditionally been assigned. Archaeologists tended to believe that people did not weave until they abandoned the migratory hunting and gathering way of life and settled into permanent agricultural villages with domesticated plants and animals, a process that was getting under way in many parts of the world by around 8000 B.C. and is known as the Neolithic. Once they were sedentary, the story went, they could develop such technologies as ceramics and weaving.

"I think this will really blow the socks off the Neolithic people because they always think they've got the first of everything," Dr. Soffer said in an interview. "We have this association of fabric and ceramics and ground stone technology with the Neolithic although we've known about ceramics from these people at Pavlov for a while, but it was written in Czech or German and it didn't make an impact."

Some scholars of the Upper Paleolithic, which in that part of the world stretches from about 40,000 to 12,000 years ago, had predicted that textiles might have been around at that time. "It's not very unexpected but it's very important," said Dr. Anthony Marks, an archaeologist at Southern Methodist University in Dallas.

Textile specialists, especially, were encouraged by the discovery.

"It indicates how important textile structures are," said Dr. John Peter Wild, an archaeologist at the University of Manchester in England. "You're way ahead of metals. The only technologies you have to compare it with in sheer brilliance of execution are stone implements. This is the organic technology that matches it."

Previously, the earliest known basketry dated to no earlier than around 13,000 years ago and the oldest piece of woven cloth was a 9,000-year-old specimen from Cayonu in southern Turkey. The oldest known twisted fibers, which could have been woven into basketry or textiles, were found in Israel and date to about 19,300 years ago.

Because baskets and textiles are made of organic materials, they perish rapidly once deposited, Dr. Adovasio said in an interview. Not surprisingly, the absence of hard evidence for textiles in the Paleolithic molded the theories on the origins and development of weaving technology.

The evidence presented in Minneapolis consists of four small fragments of fired clay bearing negative impressions of a textile or finely twined basket, Dr. Soffer said. Along with hundreds of thousands of other artifacts at the rich site, they were excavated in 1954 by Dr. Bohuslav Klima, a Moravian archaeologist. In the summer of 1990, Dr. Soffer, sorting through about 3,000 clay fragments in an effort to categorize them stylistically, noticed four pieces, about the size of a quarter, with markings on their concave sides.

She photographed them, with the notation "plant fibers?" and the next year showed them to her colleague, Dr. Adovasio, who, she said, went "absolutely ballistic."

Three radiocarbon dates of ashes at the site ranged from 24,870 to 26,980 years ago, and Dr. Soffer said the fragments could date from anytime between. She said she was entirely confident of the dating because there was no evidence at the site of any human occupation at all after 24,870 years ago, so the pieces could not have come from any other layers deposited later.

Analyzing magnified, high-resolution photographs of the fragments, Dr. Adovasio determined that two fragments bore two different weaves and two bore indistinct parallel impressions that might be from warps, the vertical threads of a weave. He could see the alignment of the plant fibrils in the photographs, so he knew the fibers were made of plant material, or bast, and not sinew, which can also be woven. Among the plants that could have provided bast were the yew and alder trees or the milkweed and nettle, the researchers said.

The archaeologists did not know whether the impressions were made intentionally or accidentally. Many of the fragments were found in ash deposits. Analysis of all four showed that they had been fired at 600 to 800 degrees Fahrenheit, which is consistent with a simple kiln or a bonfire, or even a dwelling burning down, Dr. Soffer said. One possibility is that the woven item was unintentionally pressed into wet clay near a hearth—perhaps by walking on it—and subsequently fired.

Because the fragments are so small and no selvage, or defined edge, is apparent on them, Dr. Adovasio could not determine what they came from. He said the mesh would have been similar to that in a potato sack and might have come from a bag, a mat, clothing or a basket. While it would have been possible to make the pieces without some sort of loom, it would have been far easier using one, he said, even if that meant only tying one end of the warp around a tree and the other around one's waist.

"This demonstrates an amazing investment of energy," he said.

Dr. Elizabeth J. W. Barber, a prehistoric textile scholar at Occidental College in Los Angeles, noted that plain, or true, weave involved passing a weft, or horizontal, thread over one warp thread, under the next warp thread, over the next and so on. If a nonflexible stick is woven through the warp like this, then the process can be mechanized halfway. Raising the stick lifts up every other thread of the warp (or whichever warp threads are required for the desired weave) and the weft thread can be speedily pulled through. For the following pass, the position of the separated warp threads must be reversed, and that is where a heddle, which individually holds the warp threads of the second group and attaches them to a bar, comes in.

The type of weave in the Pavlov clay fragments is "twining"; though it too can produce a cloth, it cannot be mechanized because the parallel weft threads cross each other. Dr. Barber said twining produced a more stable weave because the weft threads twisted around each other and prevented sliding.

"When you see them switching from twining over to the true weave or plain weave by around 7000 B.C., then they've figured out mechanization," she said. "They've given up stability of weave for speed of production."

Dr. Adovasio noted that twining itself was already a relatively advanced form of weaving technology. He suggested twining might even be as much as 40,000 years old.

"If they're making this, then they're making cordage," said David Hyland, an archaeologist at Gannon University in Erie, Pennsylvania. Cordage, essentially plant fibers twisted together, includes string and rope.

"And if they can make this, they can make anything in the way of a net, trap or snare," said Dr. Adovasio, who believes that because of the scarcity of evidence, prehistorians had underestimated the importance of woven materials in early peoples' lives. Conversely, he said, because of their relative abundance, stone tools have been overemphasized in archaeologists' interpretations of prehistoric economies.

"I don't buy a lot of the gender studies stuff," began Dr. Adovasio. "But mostly men have done the analysis of Paleolithic sites and they have in their minds the macho hunter of extinct megafauna. Guys who hunt woolly mammoths are not supposed to be making these."

The model of the Paleolithic men going off with spears to hunt while the women stayed home and gathered plants around the camp may be too simple, he said.

"Maybe they killed one mammoth every ten years and never stopped talking about it," Dr. Soffer said.

At the Pavlov and nearby Dolni Vestonice sites, for example, Dr. Klima unearthed far more bones of smaller animals than of mammoths. While the former may have been hunted with spears, it is more likely that nets were used to capture small animals like rabbits, the archaeologists said.

"This tool," noted Dr. Hyland, of cloth, "represents a much greater level of success where used for hunting than lithic tools."

Dr. Adovasio, who has been working with textiles for more than 25 years, said he hoped the discovery would inspire archaeologists to learn more about how textiles and basketry decayed and to pay more attention to the possibility that textiles or their impressions are preserved on sites.

One mystery is what became of the apparently advanced technologies of these central European hunters and gatherers after 22,000 years ago, when, as the weather gradually turned colder, the archaeological record of their presence in the Pavlov Hills suddenly ceased.

"You've got the huge Scandinavian ice sheet coming down from the north and glaciers coming from the Alps and you get this no-man's-land and people get out of there," Dr. Soffer said.

She suspects that some went east and some southeast. But except for a few random fired ceramics and bits of net or cord in eastern Europe, the technologies themselves remain silent for the next 7,000 to 10,000 years. When they resurface, the skills the Pavlov people employed so fancifully have been converted to practical purpose. The technique of stone grinding, instead of being used in decorative items alone, is now applied to making hoes and axes. Fired clay turns up not in figurines but in cooking and storage vessels.

"It had never dawned on these people that they could make a pot," Dr. Soffer noted.

Textiles and basketry, too, anchor themselves firmly into the technological landscape.

"It's like who invented the first flying machine? Leonardo da Vinci," Dr. Soffer said. "But Boeing didn't start making them until this century. There has to be a social and economic context for new technology. If you don't have the context, then it won't really go anywhere."

—Brenda Fowler, May 1995

# Another Ancient Innovation: City Planning

A CITY THAT THROBBED with vitality in the third millennium B.C. lies buried, forlorn and silent, beneath the windblown soil of the Upper Euphrates River Valley in southeastern Turkey. Gone are the din of commerce, the clatter on cobblestones and the cries and murmurs of family life behind mud brick walls. But the ruins speak to archaeologists of a time when a revolutionary idea may have shaped the newest cities in antiquity.

Mapping the site of the city, known today as Titris Hoyuk, archaeologists are delineating the usual urban remains. At the center once stood a palace and other public buildings on high ground. Out from there, streets ran through residential neighborhoods. Beyond city walls lay a cemetery and scattered suburbs.

On closer examination, however, archaeologists have found surprises. The streets and terrace walls appear to have been laid out and built before the houses. And the houses seem to follow a master plan, some larger than others but all of the same design.

Archaeologists are thus drawn to the conclusion that Titris Hoyuk, population 10,000 in its heyday, represents a striking example of urban planning in antiquity. Built in about 2500 B.C., this was a kind of Levittown-on-the Euphrates.

"There was a centralized vision of what a city should look like that appears remarkably similar to a typical master-planned community in the United States today," said Dr. Guillermo Algaze of the University of California at San Diego, who is directing the excavations.

Dr. Gil Stein, an archaeologist at Northwestern University who has excavated in the same region, called the explorations at Titris "very, very important research, which gives us a whole new look on what urbanism was like in the ancient Middle East."

Scholars had long ago established that the first cities anywhere arose about 5,000 years ago in the lower valley of the Tigris and Euphrates rivers, an area known as Mesopotamia that is part of present-day Iraq. Cities were presumably an outgrowth of an increasingly productive agriculture. Crop surpluses supported expanding long-distance trade and freed people to specialize in such crafts as textiles and ceramics. As the number of merchants and artisans grew, farm villages evolved into cities.

By the middle of the third millennium B.C., it now appears, the first experiments in city living were such resounding successes that people were flocking to new cities over the entire region, in what is now Syria, northern Iraq and Turkey. This excavation, and similar clues at other sites in northern Mesopotamia, suggests that only well-planned construction projects could satisfy their needs fast enough. Further work at Titris and other sites is expected to reveal some of the changing social and political forces behind this rapid expansion of urban civilization and the innovation of planned cities.

"What we're discovering is that in the third millennium a lot of very exciting and creative things are happening in all areas of the Middle East, not just the southern Mesopotamian heartland," said Dr. Charles L. Redman, an authority on early urban civilization at Arizona State University in Tempe.

Excavations at Titris, begun in 1991, are expected to be especially revealing of the form, structure and organization of these earliest cities. For one thing, Dr. Algaze and his team are concentrating research on the neighborhoods and households where ordinary people lived, not on the palaces and temples most archaeologists have focused on.

The vagaries of history also appear to have simplified the task. Titris was ultimately a failure as a city, rising and falling over a 300-year period, never again to be reoccupied. So, in contrast to most ancient cities, the ruins of Titris are not obscured by layer upon layer of debris; they lie less than three feet beneath the surface.

"I'm jealous," Dr. Stein said. " 'Archaeology' shouldn't be that easy."

Even so, archaeologists working with Dr. Algaze have so far excavated only small sections of two neighborhoods. But they have used handheld magnetometers to map the buried foundations of more than half of the 125-acre site. Outlines of the city's architecture are readily mapped because the nonmagnetic limestone foundations stand out clearly from the surrounding iron-rich soils.

By analyzing the magnetometry maps in the light of actual excavations of representative areas, the archaeologists detected patterns suggesting a planned community that, as Dr. Algaze said, "was built in one go." A study of pottery styles seemed to confirm that the houses were built and occupied within the same century.

In a report on the research, Dr. Algaze and Dr. Timothy Matney of Whitman College in Walla Walla, Washington, described excavations at the eastern end of the city that exposed at least four large multiroom houses aligned along a central street. The foundations, they said, "showed extensive evidence for urban planning in [their] regularity, plan and constructional details."

Excavations at another part of the city exposed remains of terraces built behind retaining walls that, in turn, serve as major load-bearing walls for the houses. This also suggested planning.

So did street construction. Streets examined by the archaeologists were carefully prepared, cut into virgin soil and then paved with cobbles and crushed pot shards. It was clear, the archaeologists concluded, that the streets were "laid before the houses at either side because, in places, the foundation trenches for the house walls had cut into the street."

At both excavation sites, archaeologists uncovered limestone and fieldstone foundations of several houses that were more or less identical in design. Walls rising from the foundations were made of mud bricks, which have now disintegrated. The roofs were of thatch.

One well-defined dwelling consisted of 15 rooms arranged around a central rectangular courtyard. The courtyard seemed to be subdivided with low partition walls defining what appeared to be four separate working areas. People typically entered the houses through a door off the street that led to an antechamber and then to a second small room that opened on the main house.

Such a house plan, the archaeologists said, appeared to be characteristic of domestic dwellings of ordinary people, not the elite, though some were smaller variations on the same design. The houses were probably occupied by extended families, for each one contained several cooking areas.

The houses were places of work as well as residence. In the ruins, archaeologists found several raised oval basins lined with plaster. High concentrations of tartaric acid in the plaster indicated that the basins were used in wine production, though some of the basins may have served several purposes.

Since some basins drained into the street, Dr. Virginia Badler, an archaeologist at the University of Toronto, said they would not have been used for wine but perhaps in washing and processing wool or fleece. Among the artifacts in several houses were stone loom weights, no doubt associated with weaving.

One aspect of Titris was unlike anything associated with a Levittown or other modern tract housing communities. Each house excavated so far had a family crypt, usually in the central courtyard. Though partly subterranean, the tops of the crypts were visible above the floor foundations.

Each of the tombs contained skeletons of seven to nine people, as well as the remains of food, weapons and other artifacts for use in the afterlife. One of the most poignant finds in the tomb was a perfectly preserved thistle-type flower in a vase.

The grave goods, Dr. Algaze said, promised to help establish the affluence and social status of these families in the community. Burying the dead inside private residences was not thought to be common, he said, and was not well documented in the Middle East Bronze Age until now.

Working with Dr. Algaze and Dr. Matney at Titris have been archaeologists from Britain, Italy, Turkey and the Universities of Chicago and Michigan. Dr. Lewis Somers of Geoscan Surveys Inc. conducted the magnetic mapping.

Dr. Harvey Weiss, an archaeologist at Yale University, said that similar evidence of planned cities had been showing up at other sites in northern Syria and Iraq. "We don't understand why," he said, "but this sudden urbanization and now this feature of it, urban planning, are becoming a very remarkable story of how civilization spread."

Dr. Stein said the spread of urbanization and city planning probably reflected a high degree of social organization and complexity and more powerful rulers who wanted to expand and strengthen their political bases and had the means to do it. At the time of Titris, the landscape of both southern and northern Mesopotamia was covered with competing city-states, many of which would eventually be consolidated in a single Akkadian Empire under Sargon the Great, around 2300 B.C.

The rise of cities, scholars have noted, coincided with such transforming innovations as writing, use of animals for traction, wheeled carts, metallurgy, irrigation and agriculture surpluses and craft specialization.

Dr. Redman suggested a more elemental, less materialistic reason for people's being encouraged to congregate in cities in ancient Mesopotamia and ever since. "We are a gregarious species," he said. "We like to live and work in groups."

In one sense, Dr. Redman noted, the phenomenon of the third millennium B.C. has been repeated in the last half of the twentieth century, perhaps for some of the same reasons. "We are in a period of hyperurbanization, with the Tokyos and Mexico Cities and Lagoses overflowing with humanity," he said, "because cities are the centers of learning and excitement, places for doing things and making your mark."

—JOHN NOBLE WILFORD, May 1997

# Why Live in Russia? Maybe Mammoths Were the Lure

FOR AS LONG AS HUMANS have lived here, near Moscow where the soil is poor, the winds are harsh and the winters are brutal legends—people have asked a simple, compelling question: What kind of a migratory mania could have brought people to this place?

Sure, the nomadic proclivities of the Russians have been widely noted for centuries. But why did they wander this way when, with a slight adjustment, they could have ended up in much more inviting lands to the south? Was it the first act in a many-thousand-year saga of self-loathing and defeat? Were the Russian people programmed somehow in the Stone Age to an eternally grim destiny?

Not at all. It turns out that the first residents of the Moscow area ended up here because they were in hot pursuit of mammoths, the lumbering ancestors of elephants, which flourished here during the Pleistocene epoch. The remains of more than 15 mammoths, all clearly killed by Stone Age hunters 22,000 years ago, have been discovered in Zaraysk, a small city 100 miles south of the capital.

The appearance of the hunters here, when it would have been easier for them to wander south to a better climate, "is a real revelation," said Hizri Amirkhanov, chief of Paleolithic archaeology at the Institute of Archaeology in Moscow. "We never expected them to move so far north. It changes our entire idea of the culture of mammoth hunters."

It also explains what would bring seminaked primates to a land that was even colder and less hospitable 22,000 years ago than it is today. And by stopping here rather than beyond the Ural Mountains, the mammoth hunters, who migrated from central Europe, appear to have settled a question that is debated often and remains relevant even now: Where did Eu-

Michael Rothman

rope end and the Asiatic steppes and Russia begin? This discovery shows that this region was settled by Europeans and not Asians.

"It is the most fascinating thing," said Dr. Amirkhanov, "because people are always wondering when did Europe become Europe. Why is the division where it is and not in the Urals? Well, you can start with Charles de Gaulle or with the Stone Age. But if you believe that geography is destiny, then Europe—reaching to the Urals—has been that way for at least twenty thousand years."

Dr. Amirkhanov and his team excavated this site—the oldest evidence of human life in central Russia—for more than three years. The dig actually began almost 20 years ago when residents of this quaint old Russian city noticed mammoth bones, which they first assumed were the human remains of Stalin-era executions, poking through the mud after unusually heavy spring rains. A few local historians got interested, however, and the region's significance soon became clear. But lack of money, uncertainty about what was in the mud banks here not far from the Osyotr (Sturgeon) River and confusion about how to excavate around buildings without destroying them stopped the work for years.

At first, the researchers from the Russian Academy of Sciences thought they had only found a way station for Stone Age wanderers. Flint tools unearthed in the first dig seemed to have been left by nomads. It didn't take long to disprove that theory, though. After digging nearly 20 pits, archaeologists uncovered nearly 31,000 parts of stone tools—many very small but some quite elaborate—that they say are remains of the Kostyonkovsko-Avdeyevsky culture, which had its origins in central Europe before immigrating to southern Russia over a period of at least 2,000 years.

While fundamental studies into how this community lived and why it settled here are just beginning, it has already become clear that Zaraysk was a unique prehistoric workshop. It is the only place where flint was quarried and processed to make the tools and weapons early man needed to survive. In other places, the flint was brought with the travelers.

"We know why they came," said Dr. Amirkhanov. "But there is so much more we have to learn. What were their habits and rituals? What was their community like? Did it have rules? Were there leaders here? I feel that with enough work the information at this site may help us answer some of the most basic questions about our ancestors."

One question is: How did these people kill their enormous prey? It has often been suggested that the big pits found here and elsewhere were dug to trap the mammoths. But Dr. Amirkhanov is certain that the pits were essentially prehistoric dwellings, not traps.

"You can't dig a five-meter hole in the ice for that purpose," he said. "It's not efficient."

He and others have a theory, based on the location of a few of the tools that were found, that the hunters, few of whom reached age 30, invented a system like one used today by tribes in Africa to kill elephants. "I believe they sent one or more people to crawl under the animal and stick him with spears," he said. "It was dangerous and obviously risky since it would take quite some time for the animals to die. But there was no other way that makes sense."

The site appears to contain the remains of the most developed of all European Stone Age groups. As evidence, a necklace made from the teeth of the polar fox was found here. Such decorative jewelry was unknown in earlier settlements.

Over the course of 6,000 years, at least three waves of settlers made permanent homes in the dugouts of Zaraysk, coming 21,000 years ago, 18,000 years ago and 15,000 years ago, according to carbon analysis of the residue of bones, charcoal and mammoth tusks.

The Zaraysk site—so far nearly 2,000 square meters has been excavated—was inhabited at different times and for such long periods for one reason: This place was mammoth heaven.

It is easy to understand why. The flat terrain here was perfect for a slow-moving, grass-eating mammal. Predators were easy to spot, but so was prey. There are no forests nearby, which means mammoths held a long view of the surrounding territory and lived in the midst of the high, sweeping grass they loved to eat. Not unlike the modern-day camel, the mammoth developed a hump-like reserve of fat, on the top of its head and on its shoulders, which helped it during times when the grass growth was poor. Because mammoths mostly stayed in one place, so did the early humans who thrived on them. The climate, which was then similar to today's tundra, was perfect for mammoths, though almost impossibly cold for humans.

One of the myths of mammoth extinction is that they died of the cold. Actually, the reason was far more complicated. Mammoths did not migrate seasonally because cold weather never really bothered them. They preferred to live near the Ice Age glacier that then dominated the continent. But they needed dry weather.

"High humidity and dampness is what killed them," said Lyudmila Grekhova, a senior archaeologist at the Russian State Historical Museum in Moscow, who was one of the early researchers on the Zaraysk site. "When their thick furry hair got wet and then turned to ice they would die."

The same phenomenon was recognized this century with the Siberian sable. After a new dam was built the local ecosystem warmed, and that caused higher humidity. The humidity moistened the sable's fur, and the moisture turned to ice.

Early this fall the team found two ancient dugouts that served as living quarters. "They were the first communal apartments," Dr. Amirkhanov said in jest, referring to the shared living quarters that were so common in Russia during Soviet times. They measured a little less than 100 square feet and were bolstered by intricate roofs fashioned from tusks. "They were a little bigger than a modern Soviet kitchen," he said. "But the design was better."

Dr. Amirkhanov said it is not yet known how many people would have lived at once in such a space. He hopes that further exploration, including detailed chemical analysis of the residues from nearby Stone Age campfires, will help provide an answer.

"We have about three decades of work left to do here," he said cheerfully. "In some fields that would seem like a long time. But with us it's really just a blink of the eye."

—MICHAEL SPECTER, March 1998

# 2

# EARLY CIVILIZATIONS: MESOPOTAMIA AND EGYPT

Mesopotamia, the region mostly comprising modern Iraq and Syria, is the heartland of Western civilization. The first cities were built there in the fourth millennium B.C. Records of the first writing, invented by the Sumerians of southern Mesopotamia, date from 3200 B.C. Most of what is known about the civilization comes from the work of archaeologists who have retrieved and translated the clay tablets that record the annals and accounts of this early bureaucracy.

Egypt, another ancient civilization with origins in prehistoric times, has perhaps the best-known archaeology in the world. Its dry climate has helped preserve wall paintings and papyrus manuscripts. And its elaborate funerary customs have left a wealth of relics belowground for recovery by the archaeologist's spade. Though tomb robbers long ago rifled the pharaohs' tombs, one survived untouched, and its riches proved astonishing even though its pharaoh, Tutankhamen, was one of ancient Egypt's least-significant rulers.

Although the Mesopotamian and Egyptian sites are among the most exhaustively excavated, they continue to yield discoveries. Archaeological technique, once little more than glorified treasure hunting, is continuously improving, with the result that far more information now can be recovered from a site.

# Ancient Clay Horse Is Found in Syria

A CRAFTSMAN 4,300 YEARS AGO molded the clay figurine of a horse, only five inches long and three inches high, but exquisite in detail. It has now been uncovered in northern Syria by archaeologists who call it the oldest known sculpture of a domesticated horse and one of the finest ancient representations of a horse ever discovered.

Experts said the discovery provided new evidence that the horse played an earlier, more important role in the rise of ancient empires of the Middle East than many scholars had thought.

The pale greenish figurine was found in ruins at Tell es-Sweyhat, a Euphrates River site about 200 miles northeast of Damascus, by an expedition from the University of Chicago's Oriental Institute.

Dr. Thomas Holland, an archaeologist and leader of the expedition, said the name of the ancient settlement and the identity of the people who lived there could not be determined with any certainty. The site, which could be either the ancient town of Shada or Burman, appears to have been an important trading center in the region between the empires of Akkad to the east and Ebla, a city in western Syria.

The figurine and other artifacts have been dated at 2300 B.C., about the time Sargon of Agade ruled the Akkadian Empire, successor to the Sumerians, and the region's turbulent history was being recorded in cuneiform script on the clay tablets of the Ebla library. Until recently, it was thought that domesticated horses did not figure in Middle East history until at least 500 years later.

Whoever the people were, Dr. Holland said, the care that went into making the figurine showed that the horse was highly regarded. The object may have been fashioned as a cultic figure. From the detailed representation of the genitals, it was modeled after a stallion and could have been displayed as a way of ensuring fertility for horses, much the same as figurines of women found at the site seemed intended to ensure childbirth.

A number of model chariots also found suggested that the horse was already being used to draw war chariots, as well as for transportation.

Juris Zarins, an anthropology professor at Southwest Missouri State University in Springfield and an expert on the role of horses in early Mideast civilizations, said the figurine, along with recent discoveries in the Ebla texts, showed that the domesticated horse was well established in Mesopotamia in the last half of the third millennium B.C. and contributed to the rise of the world's first large empires.

"The horse was essential to the development of empires," he said. "It made possible relatively fast transportation and empowered armies."

Determining the earliest use of the horse has long been a problem of archaeology. Recent discoveries support theories, held by Russian anthropologists, that people began riding horses at least 6,000 years ago in the nomadic societies of what is now Ukraine and southern Russia. Such an early date could mean that horseback riding was the first significant innovation in human land transport, not the invention of the wheel. Only later did the practice spread south to Mesopotamia.

Dr. Zarins said this interpretation of the earliest partnership between horse and human "is now beyond dispute." An analysis by David W. Anthony, an archaeologist at Hartwick College in Oneonta, New York, of the wear by ancient bridle bits on a 6,000-year-old stallion's tooth was crucial in settling the issue.

Archaeologists cited two clues that the figurine was of a domesticated horse. There is a hole bored through the muzzle so that a ring could be placed to hold reins. And the carefully sculpted mane lies flat, which occurs only on domesticated horses.

Dr. Holland has been excavating Tell es-Sweyhat site since the 1970s, with the recent work jointly sponsored by the University Museum of the University of Pennsylvania and the Oriental Institute at Chicago. Other discoveries, he said, included wall paintings of a large building that could have been a palace or temple, and many one-handled storage jars almost identical to ones found on Cyprus. This could be evidence that the ancient city was a link in a trading system connecting Mesopotamia with the Mediterranean.

No archive of cuneiform tablets has been found there yet, Dr. Holland said, but this could be just a matter of time. Such tablets could solve the mystery of who these people were.

It is already known, however, that their city was destroyed around 2200 B.C. Akkadian arrowheads and clay sling projectiles were found in the ruins of the mud brick town hall. Archaeologists suspect that Akkadians attacking in horse-drawn chariots overwhelmed and sacked the city where once a craftsman honored the horse with a clay figurine.

—JOHN NOBLE WILFORD, January 1993

# Lost Capital of a Fabled Kingdom Found in Syria

ONE OF THE VANISHED CITIES of antiquity, Urkesh, in its heyday more than 4,000 years ago was an opulent oasis of commerce straddling a major trade route in what is now northeastern Syria, near the Turkish border. It was reputed to be the capital of a fabled kingdom and the most sacred religious center of the Hurrians, an obscure people who were contemporaries of the Sumerians in the south and the Semites of Ebla in the west.

But history had long ago misplaced Urkesh and was left with but a dim memory of the Hurrian civilization. The Hurrians, or Horites, are mentioned briefly in the Old Testament and on a clay tablet belonging to Pharaoh Amenemhet IV, Egypt's ruler in 2000 B.C. The rest is mainly legend. Some historians had even doubted that the city ever really existed.

After a decade of excavations, an international team of archaeologists is convinced that the long-lost Urkesh has been found. They have uncovered clay tablets and seal impressions, metal tools and detailed drawings revealing that Urkesh was a real city and that its ruins lie buried beneath the modern Syrian town of Tell Mozan, 400 miles northeast of Damascus.

The archaeologists said their discoveries established that the ancient Hurrian city was more important and at least three centuries older than once thought. They were also surprised to find evidence that some women in the society appeared to own land and storehouses and to have considerable influence. Many of the seals belonged to a previously unknown queen of Urkesh named Uqnitum.

Giorgio Buccellati

**Lost Capital of Fabled Kingdom of Old Testament Found in Syria**
Drawing of seal impression of Queen Uqnitum, left, and photograph of seal impression of King Tupkish, below, found in lost city of Urkesh.

"To have succeeded in identifying it with the actual archaeological site of Tell Mozan means that Urkesh has now a geographical as well as mythical location," said Dr. Giorgio Buccellati, a professor emeritus of Near Eastern languages and cultures at the University of California at Los Angeles, who is the director of the Urkesh excavations.

Dr. Buccellati announced the discovery in Philadelphia at the joint annual meeting of the American Schools of Oriental Research and the Society for Biblical Literature. Other members of the team included his wife, Dr. Marilyn Kelly-Buccellati of California State University in Los Angeles, and archaeologists from Austria, Italy and Switzerland.

Other scholars of ancient Syrian cultures said the discovery was exciting and extremely important. It should lead to the first real understanding of Hurrian history and language, they said, and also enrich the comparative studies of art, language and cultures of other ancient Syrian societies in the period that saw the beginnings of writing, urban living, labor specialization and political empires. Hurrians seemed to occupy an upper-middle-class position in many societies throughout the ancient Middle East.

Dr. Rudy Bornaman, director of the American Schools of Oriental Research and a specialist in Syrian archaeology, called the find "a tremendous step forward."

Dr. Buccellati said that excavations at the hillside town of Tell Mozan would continue for many years. "A full assessment of early Hurrian civilization is one of the tasks which lies ahead," he said.

The archaeologists have just begun to explore the site. They have uncovered ruins of a large temple and a room described as a royal storeroom. But this represents no more than 1 percent of the 300-acre site. From 10,000 to 20,000 people once inhabited the city, archaeologists estimated. It probably flourished for several centuries in the late third millennium B.C. and then declined and faded from sight, perhaps as a result of falling water tables in an arid land.

Two bronze lions smuggled out of Syria in the 1940s and acquired by the Louvre in Paris and the Metropolitan Museum of Art in New York gave Dr. Buccellati the most crucial clue in his search. Inscribed on the base of the statuettes was writing in a strange language. Scholars deciphered the letters spelling out "Urkesh" and the name of a king who had built a temple in the city.

As Dr. Buccellati recalled, this was the exciting moment when he realized that here was evidence of a specific historical ruler, that Urkesh actually existed and that it must be buried in the vicinity of Tell Mozan. In 1987, he began excavations on the slope of the hill where farmers buried their dead, and in doing so often turned up artifacts not far from where the two bronze lions were discovered.

Acting on a hunch decades ago, the mystery novelist Agatha Christie and her husband, Sir Max Mallowan, a British archaeologist, had dug in the same area in search of Urkesh. But they gave up after two days, leaving the mystery unsolved.

The current expedition was more persistent. Its most telling discoveries were more than 600 written and drawn figures on clay seals that were found scattered on the floor of a room the archaeologists described as a royal storehouse. The seals were affixed to containers that stored goods belonging to Queen Uqnitum and some of her retainers, including her sons' nurse Zamena and her cook.

In a report summarizing the findings, the archaeologists noted that most of the seals were the queen's, not the king's, indicating that she occupied a central position in the society. "She appears as a property owner in her own right, as distinct from the king," they wrote, "and she could exercise direct control at least over that part of the storehouse where her goods were being kept."

Moreover, the name Uqnitum for the queen is Akkadian, meaning "the lapis-lazuli girl," or one who is cherished like a precious stone. The king's name, Tupkish, is from the Hurrian language. The archaeologists said this may well imply royal intermarriage between different ethnic groups.

Other drawings on clay depict events like banquets, family gatherings and a woman preparing food. One shows the king sitting on a throne with a lion crouching at his feet. Some appear to be extremely fine portrait-like representations of the queen and the prince.

Describing the discovery to other archaeologists, Dr. Buccellati exclaimed: "The Hurrians now have names, faces. We know what they looked like—we know that they lived! The crown prince has a very distinctive face, and it's not very attractive, either."

A date of 2300 to 2200 B.C. has been estimated for these artifacts, based in part on radiocarbon tests. This is two or three centuries earlier than scholars had any knowledge of Hurrian culture. But the presence of a clay tablet containing a dictionary listing names of professions seemed to confirm the age of the city. The tablet is similar to ones found at other late-third-millennium sites in Syria, especially Ebla, one of the most impressive ancient Syrian cities excavated in this century.

The archaeologists said the existence of the tablet, plus the fragments of some 40 administrative texts on clay, and the architectural layout of the building suggested that this had been a place where scribes worked.

If the excavations have indeed revealed the site of the lost Urkesh—and no one yet is disputing the claim—the discovery is expected to enable scholars to separate the ancient city of fact with the one of mythology. Kumarbi, the principal god of the Hurrian pantheon, was already known as the "father of the city Urkesh" and described as residing in Urkesh, "where he resolves with justice the lawsuits of all the lands." In mythology, Urkesh is the only known Syrian city to be mentioned as the seat of a primordial god.

—JOHN NOBLE WILFORD, November 1995

# Tomb of Ramses II's Many Sons Is Found in Egypt

RAMSES II, ONE OF ANCIENT Egypt's greatest pharaohs, is said to have fathered more than 100 children, including 52 sons. Archaeologists have discovered an enormous mausoleum with at least 67 chambers, the largest tomb ever explored in the Valley of the Kings, and they think this was the resting place for most of those royal sons.

Working their way through a narrow entrance in the limestone hillside, archaeologists were astonished to find a central hall with 16 pillars, a passageway to a statue of the god Osiris and other corridors leading away to many separate chambers. In the dim light they could glimpse wall decorations and alabaster fragments carrying inscriptions with the names of four sons of Ramses, and sarcophagus pieces, mummy fragments and statuary strongly suggesting that the tomb was used for their burials.

The sons never attained the power of their father, whose reign lasted 66 years, and their elaborate burial place is not likely to yield treasures similar to those in the tomb of Tutankhamen, the young King Tut. But the newly explored tomb, archaeologists said, promised invaluable insights into Egypt's royal family at a crucial period in ancient history.

The discovery was described by Dr. Kent R. Weeks, a professor of Egyptology at the American University in Cairo who directs a project to map the monuments at the ancient Egyptian capital of Thebes, 300 miles south of Cairo. The tomb is part of the necropolis, known as the Valley of the Kings, where monarchs and nobles were buried from 1600 B.C. to 1000 B.C. The site is on the west side of the Nile, across from the ruins of Karnak and Luxor.

"It's like no other tomb I know of anywhere in Egypt," Dr. Weeks said in a telephone interview.

Most ancient Egyptian tombs consist of only a few chambers, and these are small and laid out along a single axis. Tomb 5, as the new find is designated, not only has at least 67 chambers arrayed in a complex plan, but stairs and sloping corridors—yet to be explored—apparently lead to even more rooms on a lower level. These lower rooms may be the actual burial chambers, Dr. Weeks said, and the total number of chambers could be more than 100.

Dr. Peter Dorman, an archaeologist from the University of Chicago who works across the river at Luxor and has visited the site, praised the discovery as "a very significant find, very impressive and certain to give us a much clearer picture of the family of Ramses II." He also noted that the tomb's architecture was "completely unlike any other tomb" in Egypt.

Dr. David O'Connor, an Egyptologist at the University of Pennsylvania's University Museum, said the grandeur of the tomb complex "seems to say something important about the status of royal princes during the Ramses reign." Princes usually rated much more modest burial places.

Dr. Weeks wondered if there could be other multiple burials for the families of other pharaohs that have gone undetected.

"Tomb Five raises many questions about what else the Valley of the Kings and other areas at Thebes may have to offer," Dr. Weeks said in a statement issued by the American University of Cairo. "It is an entirely new type of New Kingdom burial structure."

Egypt and a large part of the known world in the thirteenth century B.C. were dominated by the long reign of Ramses II. He was one of the most powerful rulers in antiquity and a prolific builder of monumental architecture. His empire extended from Libya east to the valleys of the Tigris and Euphrates rivers, from Turkey south to the Sudan. His charioteers fought many battles against the Hittites, archenemies in what is now Turkey and Syria. Tradition has it that Ramses was pharaoh at the time of the Exodus of the Israelites from Egypt.

According to the Book of Exodus, the first son of the pharaoh was killed by God. In Chapter 11, Verse 5, it says, "And all the firstborn in the land of Egypt shall die, from the firstborn of Pharaoh that sitteth upon his throne, even unto the firstborn of the maidservant." The eldest son of Ramses II was named Amon-her-khepeshef. He is one of the sons whose names are inscribed on the walls of Tomb 5.

The tomb occupies a prominent place in the necropolis, 100 feet from the tomb of Ramses II himself and not far from King Tut's. In fact, when the English archaeologist Howard Carter excavated Tut's tomb, he unknowingly piled dirt and debris over Tomb 5's entrance, further obscuring it.

That entrance was discovered once in 1820 by an English traveler, but he explored no farther than the three outermost chambers, which were unprepossessing and damaged by floodwaters. In the 1980s, Dr. Weeks and other archaeologists rediscovered the hidden entrance by studying the diaries of the nineteenth-century travelers. They also followed the lead of an ancient papyrus now in a museum in Turin, Italy. It described the trial of a thief in 1150 B.C. who was caught trying to rob the tomb of Ramses II and also one "across the path," presumably the one now called Tomb 5.

Cutting test trenches on the slopes near the Ramses tomb, they uncovered the narrow entrance and did some preliminary investigation. But debris washed in by floods hampered exploration. Only in February were Dr. Weeks and his team able to remove debris blocking the door leading off the grand hall with 16 pillars.

Once through the door, they found less water damage and one surprise after another. They followed the passageway leading past 20 doors of smaller chambers and ending at the statue of Osiris, god of the underworld. Two branching corridors each had 20 doors to other rooms. The smallest rooms were about 10 by 10 feet; the largest was 60 by 60 feet.

This was when Dr. Weeks realized he had entered, as he said, "the largest tomb in the Valley of the Kings and maybe the largest ever found in all Egypt." It was certainly more complex and extensive than the standard royal tombs, which are simple structures with a single corridor leading to a main burial chamber.

In an interview with the Reuters news agency in Cairo, Abdelhalim Nourredin, head of Egypt's Supreme Council for Antiquities, described the structure and the many chambers as magnificent, even if some of the rooms were badly damaged by water, and important because of its "unique design and size from a crucial period in ancient history." The Egyptian government worked with Dr. Weeks in the explorations.

Dr. Weeks speculated that the rooms on this main level were used for religious ceremonies and making offerings to the dead, with the lower rooms set aside for the burials.

The floors were littered with thousands of pieces of pottery, statue fragments, jewelry and beads, pieces of wooden furniture, stone sarcophagus fragments, inscribed stone vessels, bones from offerings of cooked meat and pieces of mummified human bodies. Dr. Weeks said there was no indication that looters had ever penetrated the back chambers, or been anywhere in the tomb since antiquity.

Close examination revealed the inscribed names of four sons of Ramses. Decorations in the small entrance rooms included the names of Amon-her-khepeshef, the firstborn, and the second son whose name was, in effect, Ramses Junior. A piece of alabaster jar bore the name and titles of another son, Sety, and outside the entrance archaeologists found a piece of limestone on which was written the name of Mery-amon, the fifteenth son.

Historians know the names of all 52 sons, but little else about them. They had previously been able to identify the burial sites of only two of Ramses' sons. Mernepteh, the thirteenth son, succeeded his father as pharaoh and so built his own tomb in the Valley of the Kings. The fifth son, Khaemwase, is thought to be buried at Saqqara, south of the Pyramids of Giza.

It is entirely possible, Dr. Weeks and other archaeologists said, that all 50 of Ramses' other sons were laid to rest in Tomb 5. Explorations are scheduled to be resumed in July.

—JOHN NOBLE WILFORD, May 1994

# Long-Lost Field Notes Help Decode Treasure

FROM THE OUTBREAK OF World War II until the end of the Cold War, several thousand artifacts of ancient Egypt lay in Chicago and Cairo museums, a treasure haunting and intriguing yet beyond scholarly interpretation. The pieces were like cryptic messages from the past whose code books had been lost to archaeologists.

The statues, jewelry, objects of erotica and various religious symbols were excavated in the 1920s and 1930s by American and German archaeologists working at Medinet Habu, a major site of temple ruins on the west bank of the Nile River at ancient Thebes, known today as Luxor. As is the usual practice, the archaeologists kept meticulous notes of exactly where each relic was found—in which buildings, at what depths in the sediments and next to what other artifacts. Without such records, archaeology is little more than a treasure hunt, and the materials cannot be reliably dated or their function and meaning reasonably surmised.

In 1939, when war erupted in Europe, these field notes, in 10 volumes of more than 1,200 pages, were in Berlin being prepared for publication. They disappeared in the confusion of war and remained lost in the Cold War isolation of East Berlin, where they had last been seen. For all anyone knew, the Medinet Habu field notes had been destroyed.

Then, with the collapse of the Soviet Union and the reunification of Germany in 1990, a letter from Berlin museum officials arrived at the Oriental Institute of the University of Chicago. All the notes had been found and would be reunited with the artifacts.

"It's a wonderful dividend of the end of the Cold War," said Dr. Emily Teeter, an assistant curator of Egyptology at the Oriental Institute, who returned from Berlin with the long-lost records. "The rediscovery of the field notes makes it possible to incorporate the objects into the history of Egypt."

Egyptologists said that reexamination of the relics in light of the field notes should set in motion a new wave of research not only into the elite culture of ancient Egypt but also into the everyday lives of common people under the pharaohs and up until Roman and Christian times, a subject of increasing importance to scholars of this early civilization. Most previous studies have concentrated on Egyptian royalty.

As a sacred royal center, Medinet Habu was the site of the temple of Pharaoh Ramses III, who reigned from 1182 to 1151 B.C. This is one of the largest and best-preserved mortuary temples of ancient Egypt. Standing at the foot of a mountain, its walls are elaborately decorated with hieroglyphs and scenes of battles with Nubians, Libyans and the enigmatic Sea People. Other temples perpetuate the memory of Aye (1325–1321) and Horemheb (1321–1293), who succeeded Tutankhamen. Queen Hatshepsut (1503–1483) erected a temple there dedicated to Ogdoad, four pairs of deities who in Egyptian mythology were responsible for the creation of all people.

Beginning in the time of Ramses III, Medinet Habu became more than a religious center, expanding as an administrative center for western Thebes with government offices and warehouses and a growing population living in private homes all surrounded by city walls 60 feet high. The 5,000 artifacts excavated there represent one of the largest collections of materials from a single site in Egypt and include remains from every period of the site's occupation, from about 1500 B.C. until A.D. 800.

But until the notebooks were recovered, Egyptologists despaired of ever being sure of the significance of many artifacts, especially those apparently related to religious practices, because they did not know which objects came from temples and which from private houses.

"By determining which objects came from commoners' homes, we gain whole new insights into the Egyptian way of life," Dr. Teeter said. "This material allows us to see the art and ritual objects of the common man and woman and to see how most of the society lived."

Other Egyptologists were also ecstatic. "It's so very important to our research," said Dr. Dorothea Arnold, curator of Egyptian art at the Metropolitan Museum of Art in New York City. Dr. Richard Fazzini, curator of

Egyptian, classical and ancient Near Eastern art at the Brooklyn Museum, said the notes should pinpoint the dates of the most interesting art and ritual artifacts and help establish their purpose. "This means the materials will no longer exist in a vacuum," he said.

The dig at Medinet Habu from 1927 to 1933 was the biggest the Oriental Institute ever conducted in Egypt. The expedition was directed by Dr. James H. Breasted, a professor at the University of Chicago who was one of the preeminent archaeologists of the day, and most of the excavating was done by an experienced German team led by Dr. Uvo Holscher. Besides exposing the grandeur of the temples and tombs, the archaeologists collected more than 5,000 objects, including statuary, clay figurines, glazed plaques, tools, weapons, offering tables, pottery, scarabs and amulets.

These artifacts were divided between the Oriental Museum and the Cairo Museum, but the field notes went to Berlin, where they were to be published by Dr. Rudolf Anthes, an Egyptologist. When he fled Berlin during the war, he left the notebooks at the Bode Museum. After the war, he wrote to Dr. Holscher saying that the notebooks might have been destroyed by Allied bombing or seized by the Russians, in which case they were "probably lying somewhere in Russia and are rotting there."

Scholars at the University of Chicago said they had not been told where the notebooks were found or where they had been for the last 50 years or more. All they know is that two years ago, officials of the Bode Museum in the former East Berlin notified the university of the reappearance of the notebooks. The Germans offered to give them to the Oriental Institute, which they did.

Poring over the documents, Dr. Teeter discovered that they were in excellent condition, beautifully written in various colors of ink and by several different hands. The pages are crammed with notes and the occasional rubbing of an inscription. Sometimes conclusions written one day had been amended later by others, using different-colored ink, as new evidence was uncovered.

One of the first mysteries Dr. Teeter sought to solve with the notebooks involved the many objects found at Medinet Habu that archaeologists call votive beds. Made of baked clay, each is about the size of a tall, narrow doll's bed about a foot long and is decorated with religious symbols. Scholars assumed these small beds were used in fertility rituals.

From the Berlin notebooks, Dr. Teeter and her colleagues discovered that the beds had been found accompanied by clay female figurines and came from private houses, not temples. It was presumably the first of many new insights scholars expect from the notebooks. The beds and figurines are examples of folk art and folk religion, in contrast to the more formal religious rites and symbols of the temples.

"Now we are sure that the votive beds are a reflection of an ancient fertility cult enacted in private homes," Dr. Teeter said. "The material is an intriguing reflection of the cares of women and their families in ancient Egypt."

Erotic art in the collections, mostly figurines of women with extremely large breasts and men with highly exaggerated phalluses, also appeared to be related to fertility rites of folk religion, the notebooks indicated.

Since the notes include stratigraphic data of the excavations, the Chicago scholars have been able to establish the continuity of practices related to the worship of Osiris, a major god of the afterlife. In the collection there are hundreds of bronze statues of the god that were excavated from the Osiris "grave" near the Ramses III temple, but it was not clear whether they were deposited there in one brief period or over centuries.

The notes showed that the statues had been found in three distinct layers, Dr. Teeter said, suggesting that the practice of making these offerings to Osiris had continued for several hundred years.

Scholars have yet to pluck from the notes any clues to the meaning of a small clay head of a man, which seems to show that the exalted rulers of ancient Egypt were not immune to derision. The clay figure is an apparent caricature of a pharaoh, for it depicts a man with a crown and a ridiculously enlarged nose.

After only a few days of analysis, the notebooks so far have produced no big surprises, scholars concede, but these may come in the years of research that lie ahead. Of most importance, Dr. Teeter said, "we can start to publish material about Medinet Habu with confidence we have the full story."

The notebooks will be available to scholars from other institutions, and eventually their contents will be published.

The end of the Cold War could lead to the recovery and repatriation of other prizes of archaeology. Also from the former East Berlin, for the first time since the war, will come a rich collection of gold jewelry from the kingdom of Meroe in the Sudan during the second millennium B.C. to be exhibited at the Metropolitan Museum of Art in New York.

Even the most celebrated archaeological treasure reported missing in World War II, the golden relics of ancient Troy, should soon see the light of day. Last month museum officials in Moscow said what had long been suspected: Soviet troops had seized the so-called Treasure of Priam in the battle of Berlin and shipped it to Russia, where it disappeared in a secret vault in Moscow.

This rich hoard of gold cups, jewelry and artifacts, some 12,000 pieces in all, was excavated by the amateur German archaeologist Heinrich Schliemann in 1873 when he discovered the buried ruins of Troy, the city of Homeric legend. Even though scholars now dispute Schliemann's conclusion that the gold had belonged to King Priam, who ruled Troy around 1200 B.C. and is described in Homer's *Iliad,* this has not tarnished the treasure's allure.

But the rediscovery of the Trojan gold, perhaps because it is gold, has provoked a rush of covetousness—nothing like the amicable return of the Egyptian field notes. Turkey claims ownership of the treasure because it was excavated from Turkish territory. Germany's claim is based on the fact that Schliemann willed the gold to "the German people" in perpetuity. No matter, the first exhibition of the treasure, Russian museum officials say, is to be at the Pushkin Museum of Fine Arts in Moscow after the long-hidden collection is inventoried and studied by world experts on ancient Troy.

—JOHN NOBLE WILFORD, September 1993

## Long-Lost Notebooks of Egyptian Expedition

More field notebooks written during a major archaeological excavation in Egypt, long feared to have been destroyed in World War II, have turned up in Germany.

The recovery of the daily excavation logs and artifact lists, with other materials discovered earlier, should enable archaeologists to complete their interpretation of the tombs, temples and other sacred remains at Medinet Habu.

News of the rediscovery of the first batch of notebooks sent an archaeologist's grandson into the family attic, where he found the rest of the missing notebooks.

The University of Chicago announced that its Oriental Institute had received the second set of valuable records, which fill 11 volumes, from descendants of Dr. Uvo Holscher, the German archaeologist who was the expedition's field director. Artifacts from the excavations had been divided between the Oriental Institute Museum in Chicago and the Cairo Museum. But the field notes had gone to Berlin, where they were being prepared for publication when war broke out in 1939 and had not been seen again.

"It's extraordinarily important and very exciting," said Dr. Emily Teeter, an assistant curator of Egyptology at the Oriental Institute. "Now we will be able to complete our documentation of the site and its thousands of artifacts."

The university learned of its windfall in a letter earlier this year from the late field director's grandson, also named Uvo Holscher. When he read in the *New York Times* about the return of the first set of Medinet Habu notebooks, the younger Dr. Holscher explained, he recalled the old documents of his grandfather that used to be stored in the family attic in Hanover, Germany.

The notebooks were found at the Technical University of Hanover during restoration work in 1972, nine years after the elder Dr. Holscher's death. He had been a professor there. At the time, his widow wrote the Oriental Institute about the find, but apparently the letter was never received, and the Chicago scholars continued to assume that all of the expedition records had been lost.

But on his next visit to Hanover, the younger Dr. Holscher went into the attic and there the notebooks were, as well as about 1,000 photographic negatives from the expedition. Four volumes were the handwritten daily log kept by his grandfather. The seven other volumes were lists of excavated artifacts, including many detailed site plans and architectural drawings.

These notebooks are duplicates of material found earlier in the Berlin museum, Dr. Teeter said, but they are complete and fill in an important gap in expedition records for excavations conducted in 1929, 1930 and part of 1931. The volumes include much anecdotal material about daily expedition life, from the travail of transporting the colossal 17-foot-high statue of King Tutankhamen a half mile in four days to the hiring and firing of workers and details about exploring the tomb of Amonirdis, the daughter of a Nubian pharaoh in the eighth century B.C.

From what is already known, Medinet Habu was a major religious center as well as a town from the twelfth century B.C. to the eighth century A.D. Dr. Teeter said the newly recovered notebooks completed the expedition's full set of records.

—JOHN NOBLE WILFORD, May 1994

# Forgotten Riches of King Tut: His Wardrobe

WHEN THE EGYPTOLOGIST Howard Carter uncovered the tomb of the Pharaoh Tutankhamen in 1922, it was the brilliant gold of the funeral mask and other artifacts that awed the world. But the tomb also contained wooden chests filled with the boy king's clothes and footwear.

Along with most of the rest of the treasures, the bulk of the textiles, some reduced to dust, ended up in a storeroom at the Cairo Museum. When Mr. Carter died in 1939, his Tutankhamen archive, including 1,500 photographs, many drawings and 2,500 note cards documenting the textiles, was deposited at the Griffith Institute at Oxford University in England. And there the materials sat for more than five decades, virtually forgotten by all but a few scholars whose general opinion was that nothing could be done because of the supposed poor state of the textiles and the difficulty in obtaining access from the cautious Egyptian authorities.

But in 1991 a friend urged Dr. Gillian Vogelsang-Eastwood, a textile archaeologist in the Netherlands, to take a look at Mr. Carter's notes. At the time she was writing a book on Egyptian clothing and how to identify it. When she began examining the material, she was incredulous. Here was outstanding documentation of the only surviving clothes of the most famous Egyptian king and yet nobody was working on it.

"I went out and demanded a cup of tea," she recalled in a phone interview. "Then I walked around Oxford for a couple of hours thinking, 'No, I'm not going to do this.' "

But she is doing it. She and her students at the Stichting Textile Research Center at the National Museum of Ethnology at Leiden in the Netherlands have identified and catalogued about 80 percent of the surviving wardrobe of Tutankhamen, the boy who became the ruler of Egypt at the age of nine and died unexpectedly and inexplicably just nine years later, in about 1324 B.C.

Among the many textiles are 145 loincloths, 12 tunics, 28 gloves, about 24 shawls, 15 sashes, 25 head coverings and four socks, which had separate places for the big toe so that they could be worn with the 100 sandals, some worked in gold. There are also one golden and one beaded apron, real leopard skins and even one faux leopard skin woven of linen with appliqu d spots. The tomb also contained a belt and tail of gold and lapis lazuli and sleeves with wing-like flaps that Dr. Vogelsang-Eastwood thinks were worn to imitate the wings of gods and goddesses.

"The project itself is of tremendous importance," said Dr. Emily Teeter, an Egyptologist and assistant curator at the Oriental Institute Museum at the University of Chicago. "The Tutankhamen clothing is really the only major group of royal clothing we have. It is one thing to look at the many paintings of these people, but knowing the type of the fabric, how the clothing hung and the colors is going to tell us a tremendous amount about how these people looked."

Rosalind Janssen, an assistant curator at the Petrie Museum at University College London, said the main importance of the work was that it would provide a counterpoint to what was already known about the wardrobes of ordinary people in ancient Egypt.

Early in the project Dr. Vogelsang-Eastwood went to the Cairo Museum to try to see the king's garments, a few of which had been on display there for decades. After earnest discussions with her Egyptian colleagues, she was led into a storeroom filled with chests and boxes of the clothing. There, she said, she found many of the clothes still in the boxes in which Mr. Carter had placed them. A few fragments of clothing lay on a copy of the *Egyptian Gazette*, an English-language newspaper dated December 22, 1922, six weeks after the discovery of the tomb.

Though encouraged by the state of preservation of most of the clothes, she is also distressed by their worsening condition. Some had decayed considerably by the time Mr. Carter discovered them. The weight of the gold and beads on the clothing had torn some apart; other textiles were probably damaged by the messy repacking done by the ancient necropolis guards after the tomb was twice penetrated by robbers shortly after Tutankhamen's death. Primarily because of a lack of money and other priorities at the Cairo Museum, the textiles have not yet been properly conserved.

"I'm scared that if nothing happens soon they'll be gone in fifteen years," said Dr. Vogelsang-Eastwood, who is trying to raise money for a textile conservation laboratory at the museum.

Her main concern is to get the textiles into climate-controlled storage. Restoration is not possible because the fabrics are too delicate; moreover, such restoration is now considered too aggressive. But, using ancient methods, expert weavers at the Hand-Weaving School in Boras, Sweden, are planning to make exact reproductions of about 20 garments, including some with gold. Since the originals can never be transported for display, Dr. Vogelsang-Eastwood expects the reproductions to form part of an exhibit of the king's clothes planned for the year 2000.

But the first reproductions were made by Dr. Vogelsang-Eastwood and her students, who found that only by trying the garments on and wearing them about could they understand how the Egyptians wore them. Mr. Carter had described one particular tubular garment with odd flaps as a hat, but Dr. Vogelsang-Eastwood could not make it stay on her head. She eventually discovered that it fit onto her upper arm like a sleeve. Worn like that, the flaps looked like wings. Now she believes that Tutankhamen wore the sleeves to imitate certain winged Egyptian gods.

Egyptian clothing had no hooks; tucking, wrapping and tying were the sole means of keeping the clothes on the body.

"People were constantly adjusting their clothing," said Dr. Vogelsang-Eastwood. "You learn to take small steps and walk in a more fluid manner."

In addition to providing insight into ancient Egyptian weaving technology and sartorial customs, the cataloguing work is illuminating aspects of this prosperous period of Egyptian history, like social and international relations.

From her study of the originals in Cairo, Dr. Vogelsang-Eastwood determined that the general structure of clothes was identical for royalty and commoners. The most basic item of clothing at the time was the loincloth, a long piece of linen shaped like an isosceles triangle with strings coming off the long ends that was worn by both men and women. Mr. Carter found many of Tutankhamen's loincloths wrapped in bunches of a dozen. The garment was tied around the hips with the material hanging down the back. That was then pulled through the legs and tucked over the string from the outside in.

Tutankhamen's loincloths, which probably served much like modern underwear, had the same structure as those of his subjects, but the quality of the material and the tailoring differed greatly. The linen in the loincloth of an ordinary Egyptian had 37 to 50 threads per inch, while the linen in Tutankhamen's loincloths had 200 threads per inch. The linen was hand-woven from threads of just three or four filaments of flax, a weave so fine it feels like silk, Dr. Vogelsang-Eastwood said.

Whether the pharaoh had in fact worn the clothes entombed with him during his brief life was another question Dr. Vogelsang-Eastwood had when she began the research. Analyzing the loincloths with a hand lens, she noticed stresses in the weave that indicated the garments had indeed been worn. Dr. W. D. Cooke, a textile technologist at the University of Manchester Institute of Science and Technology in England who was one of the project's many collaborators, examined a piece of one tunic with a scanning electron microscope and found frayed, fibrous threads. He concluded that the tunic had either been washed about 40 times in water or been washed less frequently in natron, a solution of sodium carbonate that whitens as it cleans. Unlike the clothing of the people, the king's clothes are not full of mends. Dr. Vogelsang-Eastwood has found only one tear, in a section of a sash that would have been subject to pulling.

Over the loincloths Egyptians wore tunics, simply shaped garments made by folding a length of cloth in half, sewing up the sides, leaving room for armholes and cutting an opening at the seam for the head. In Tutankhamen's day, men wore either short ones that reached just below the buttocks or long ones that fell below the knee, while women wore only the longer ones.

Among the 12 tunics found in the tomb are two decorated with blue faience beads and gold disks and two with stripes.

"Some of the garments are just so heavy with beadwork and gold, I can't see the lad wearing them every day," Dr. Vogelsang-Eastwood said. "Some are only decorated on the front so he could sit."

One particularly beautiful tunic, which has hung on display in the Cairo Museum for decades and was mistaken for a rug, was woven using a tapestry technique, in which threads of certain colors are added at just the point where they are desired for the design. To some early observers the design appeared to be printed or painted on because the weave, about 225 threads to the inch, is so fine. Dr. Vogelsang-Eastwood believes that weaving the tunic, whose chest section is decorated with a cartouche containing the name Tutankhamen as well as a line of previously unrecognized inscriptions, would have taken several months.

Another closely studied tunic is one whose embroidered iconography, including palmettes, had long been held to be Syrian or Syrian influenced. Yet it also had Egyptian motifs and the king's name appeared twice, once embroidered onto the tunic and once woven into it. Because none of the sewing techniques found in the tunic show up in other Egyptian clothing, Dr. Vogelsang-Eastwood now thinks the entire tunic was constructed in Syria by many different experts and presented by the Syrian court to its Egyptian counterpart.

Dr. Teeter noted that there are letters starting even before Tutankhamen's reign referring to trade between the Babylonian, Mesopotamian, Syrian and Egyptian courts. "Some of these records are actually one king complaining to the other king, saying, 'You're not sending good enough stuff,' " she said.

Among the many textile fragments is one in a weave no one had ever identified before. The weave is a type of sheer tapestry with a zigzag pattern in red, blue and white.

"It was like finding a new language," said Dr. Vogelsang-Eastwood, who turned to textile history and archaeology after studying to become an embroiderer in her native England.

The next phase of the project will partly build on the experience of wearing these clothes. "We've identified what the items were, and now we've got to work out when he wore it and why," she said.

—BRENDA FOWLER, July 1995

# 3

# THE CLASSICAL WORLD

Western civilization traces its roots to Greece and Rome, two predecessors that have exerted an overwhelming influence on its language, culture and institutions. A significant fraction of Greek and Roman literature has survived, giving historians a deep understanding of how these societies worked.

Still, archaeology has made significant contributions, providing firmer dates for some historical events and illuminating many of the circumstances of everyday life that never found their way into written sources.

Archaeologists are on strongest ground when their evidence is corroborated by written sources. Thucydides, the ancient Greek writer who chronicled the great war between Sparta and Athens, notes in a prophetic passage of his history that if ever the two cities should be reduced to ruins, with only their temples and the foundations of their public buildings left standing, the spectator would judge Sparta to have been far less powerful than it was. Indeed, the temples of the Parthenon still draw visitors from all over the world, whereas of Sparta scarcely two stones remain standing.

# Santorini Volcano Ash, Traced Afar, Gives a Date of 1623 B.C.

ASH BELIEVED TO BE FROM a great explosive eruption that buried the Minoan colony on the island of Santorini 36 centuries ago has been extracted from deep in an ice core retrieved in 1993 from central Greenland. Its depth in the core indicated that the Aegean eruption, which may have given rise to the Atlantis legend, occurred in or about 1623 B.C.

From the top half of the 9,000-foot core evidence has been found of some 400 volcanic eruptions in the past 7,000 years. The ash spewed into the air was high and voluminous enough to reach Greenland, about 3,500 miles away. A prominent ash layer at a depth corresponding to 4803 B.C. may have come from the eruption in Oregon that destroyed Mount Mazama, leaving the giant caldera that is now Crater Lake.

Results of the analysis were reported in the journal *Science* by Dr. Gregory A. Zielinski of the University of New Hampshire and colleagues at the university and from the army's Cold Regions Research and Engineering Laboratory in Hanover, New Hampshire, and Pennsylvania State University.

The study was part of the Greenland Ice Sheet Project 2, which extracted an ice core from the entire thickness of ice at Greenland's summit. A second core extracted nearby by a European team is also being analyzed.

Dr. Zielinski has taken microscopic ash fragments from some of the largest eruptions, including the one believed to have occurred at Santorini, to Queens University in Belfast, Northern Ireland, for analysis. Chemical analysis of ash from the eastern Mediterranean and Black Sea has shown that it all apparently came from the Santorini explosion.

Because wind systems in the Northern and Southern Hemispheres are somewhat independent, most eruptions evident in the Greenland ice have been attributed to volcanoes in the Northern Hemisphere. But there are exceptions. One in about A.D. 177 is believed to have been at Taupo, New Zealand, whose ash may have risen almost 40 miles.

Ash layers in the core have been identified by their sulfur content. Fifty-seven of 69 events recorded for the last 2,000 years were matched with known eruptions. This was true, however, of only 30 percent of the older record, to 7,000 B.C.

The Greenland core records 18 huge eruptions that took place from 7,000 to 9,000 years ago, depositing unusually heavy layers of ash. That was when the great ice sheets were melting and, the authors of the *Science* article suggest, may have been when molten material deep within the earth's volcanic zones welled up in response to the diminishing burden of ice. Those zones included Kamchatka, the Aleutians and Iceland, all upwind of Greenland or relatively near.

The earliest exactly dated eruption was that of Vesuvius, which destroyed Pompeii and Herculaneum in A.D. 79, preserving their precious frescoes under a blanket of ash. The same thing happened 16 centuries earlier at Santorini, which is also known as Thira. The island was buried under ash that in places was more than 900 feet deep, preserving wall paintings that document in vivid detail the Minoan way of life.

Wall paintings on Crete, the chief Minoan center 75 miles to the south, were not similarly protected from weathering, earthquakes and tidal waves and have been a major restoration challenge.

Ash from the Santorini explosion has already been identified deep in sediment layers on the floor of the eastern Mediterranean, in Egypt's Nile Delta and in parts of the Black Sea. There are also suspicions that its ash cloud persisted long enough to stunt the growth of oak trees in Irish bogs and of bristlecone pines in the White Mountains of California, producing tightly packed tree rings.

Uncovering the buried city on Santorini was first stimulated in the 1860s when it was found that the ash made ideal waterproof cement. Shiploads were exported to build the Suez Canal, but not until 1967 did large-scale excavation of the buried city begin, to be led for many years by Dr. Spyridon Marinatos of Greece.

The demise of the Minoan civilization has long been a mystery and for many years Dr. Marinatos attributed it to ash clouds, earthquakes and tidal waves from the Santorini eruption and the collapse that formed its caldera. More precise datings, however, indicate that the Minoan decline on Crete came many years later.

The eruption, however, was clearly catastrophic and many archaeologists believe that flooding and burial of Akrotiri, the Santorini city, could have been the basis for Plato's account of Atlantis. Layering in walls of the Santorini caldera show that it has been the scene of many catastrophic eruptions.

Plato's account is the primary source of the Atlantis legend. He attributed the account to Solon, an Athenian statesman of an earlier century. Many elements of the story seem improbable, such as an attack on Greece 9,000 years earlier by warriors from an island, "Atlantis," in an ocean beyond the Pillars of Hercules (the Strait of Gibraltar). Yet Plato's description of the destroyed island refers to many features, like the pursuit and sacrifice of sacred bulls, that were hallmarks of the Minoan civilization of Crete and Santorini.

The Atlantis invaders, said Plato, were defeated when there were "violent earthquakes and floods; and in a single day and night of misfortune all your war-like body of men in a body sank into the earth, and the island of Atlantis in like manner disappeared in the depths of the sea."

—WALTER SULLIVAN, June 1994

# Outer "Wall" of Troy Now Appears to Be a Ditch

WHEN ARCHAEOLOGISTS RESUMED excavations at ancient Troy, they expected to uncover remains of a thick wall that marked the outer limits of the city at the time of the legendary Trojan War in the thirteenth century B.C. Discovery of the wall, which they thought they had detected the previous summer, would for the first time define the full size of the city, well beyond the inner citadel excavated in the 1870s.

Unless their geomagnetic survey was mistaken, archaeologists thought buried remnants of the wall would be found about 1,300 feet outside the central fortress and its palaces, where the wall presumably encompassed the much-larger settlement of craftsmen, merchants and soldiers and sailors. The archaeologists dug and dug in the area where the survey had indicated the presence of a thick clay wall, but could find no sign of it.

Instead, they found the next best thing: evidence of a wide ditch cut into the bedrock, almost certainly an obstacle against invaders. Archaeologists said the trench probably served as the first line of defense on the city's southern perimeter. A high wall may have stood inside the line of the trench, they said, but its stones were long ago removed and reused.

Dr. Manfred Korfmann, an archaeologist at the University of T bingen in Germany and director of the excavations, said the findings established that ancient Troy was indeed one of the largest known cities in the area of the Aegean Sea during the late Bronze Age. The ruins dug up in the 1870s by Heinrich Schliemann, the gifted German amateur archaeologist, were 600 feet in diameter, seemingly too modest for a city of Troy's supposed wealth and power, as characterized in Homer's epic of the war between the Greeks and Trojans.

"When we didn't find a wall, we were disappointed at first," Dr. Korfmann said in a telephone interview. "We continued exploring, going deeper, and then we came upon this ditch. It's just as important, because it answers our questions about the city's actual size."

Dr. Korfmann said the excavated parts of the ditch were about three feet deep and 13 feet wide, presumably large enough to impede the movement of invaders with battering rams and other instruments for breaching a city's gates and walls. Further digging should determine if the ditch encompassed the entire city or only the south side.

The ditch was found just where the geomagnetic survey had indicated there was an underground anomaly, a change in soil density that archaeologists had interpreted as the possible remains of the city wall. It was an understandable mistake. Since the ditch had become filled with tile and pottery fragments and other refuse, it would have produced magnetic readings similar to density variations caused by a buried wall.

Dr. C. Brian Rose, an archaeologist directing the University of Cincinnati's participation in the excavation, said that reconnaissance of the plain between the citadel and the outer trench has revealed considerable evidence of large numbers of people living there in the late Bronze Age. The area was also occupied in Roman times, when the city was called Ilion.

Dr. Rose's team, concentrating on the Roman history of Troy, reported finding a larger-than-life-sized marble statue of Emperor Hadrian, who ruled the Roman Empire from about A.D. 117 to 138. The statue was found behind a stage in a Roman theater.

—JOHN NOBLE WILFORD, September 1993

## Third Defense Line Found at Troy

In their continuing explorations of ancient Troy, archaeologists have uncovered evidence of a wooden palisade that served as an additional line of defense beyond the stone walls of the city in the late Bronze Age.

Archaeologists also reported the discovery of a marble head of Augustus, the first emperor of Rome, in the ruins of the Odeion theater near the center of Troy, which is in western Turkey.

Dr. C. Brian Rose said the bust was slightly larger than life and in almost perfect condition. Two graduate students on the team, Cem Aslan and William Aylwar, made the discovery.

Dr. Rose, an authority on Roman imperial portraiture, recognized the image of Augustus immediately by the hairstyle, shape of the nose and furrowed brow. Augustus is a familiar sight in classical archaeology because many busts of him have been found scattered around the Mediterranean.

Finding an Augustus bust at Troy was not surprising, Dr. Rose said in an interview, because it was already known that the emperor visited the city in 20 B.C. and that he, like many elite Romans, said he was descended from ancient Trojans. Such an ancestry was traced back to Aeneas, who supposedly led a band of refugees out of the burning city after the legendary war in about 1200 B.C. that was chronicled by Homer. These refugees were thought to be the forerunners of the Roman people.

Dr. Rose said the discovery provided further evidence that the Odeion, one of the main attractions for visitors today at the site of Troy, had been erected during the reign of Augustus, when the city was Roman and called Ilion.

Of greater significance is the discovery by Dr. Manfred Korfmann's team of traces of extensive wooden fortifications at Troy.

The new discovery revealed parallel lines of postholes and other evidence of a wooden palisade that stood about 500 feet south of the stone walls, within the perimeter of the newly discovered trench.

The stone, wood and trench barriers, Dr. Rose said, "certainly demonstrate the sophistication of Troy's defensive system during the late Bronze Age," which ended about 1000 B.C. But the wooden palisade, he cautioned, cannot yet be linked to the Homeric Trojan War and does not prove that it actually took place.

—JOHN NOBLE WILFORD, September 1997

# Ancient City of Petra Is Yielding Its Secrets

THE 2,000-YEAR-OLD TOMBS of Petra, streaked with white- and rose-colored Cambrian sandstone, have long been known to border the ruins of one of the greatest ancient cities in the Middle East.

But the metropolis of Petra, unlike Palmyra in Syria, has never been extensively excavated, and visitors might have had the impression that the site was little more than a giant necropolis.

Now an international team of archaeologists has begun an extensive project to uncover the lost city, which played an important role in the Nabatean Empire, which dates from approximately the third century B.C. to the second century A.D., and in the Roman and Byzantine empires, until its virtual abandonment at the end of the sixth century. Work has begun on a Byzantine church discovered in 1990, but the rest of the city still lies under sand dunes.

"Most of the city structures have been covered by naturally deposited sand," said Dr. Zbigniew T. Fiema, who is working on the excavation of the church as part of a team from the American Center for Oriental Research in Amman. He stood under a tarpaulin near the work site at the church. "The winds are very strong in this area, so I expect, in many areas, we will find magnificently preserved structures up to one or two stories high," he said. When the winds are strong, structures are buried more quickly by the sand.

Dr. Fiema's excavations at the church have already produced at least one potentially major find. Recently archaeologists announced the discovery of scores of papyrus scrolls that are more than 1,400 years old and could date back to Roman times. Deciphering the scrolls will be extremely difficult, though, because they were carbonized in a fire that swept through the church when it collapsed in an earthquake, perhaps in the year 551.

"Only when these carbonized papyri are separated, conserved and carefully unwrapped can the script be identified and understood," a representative of the American Center for Oriental Research said. The international team also includes the University of Utah, Brown University and Swiss and French teams.

Archaeologists hope that excavations in Petra, which could have had as many as 30,000 inhabitants, will eventually help fill in many of the mysteries of the Nabatean civilization, including an understanding of its pagan beliefs, daily life and cultic practices. The Nabatean kingdom is considered to have been an important force in the ancient world, but it is one often neglected by modern scholars.

"The city of Petra has hardly been touched," said Dr. Pierre M. Bikai, the director of the American center, "and we expect to find many surprises. This is a first-class site, and one of the great archaeological treasures in the Middle East."

Petra is famous for its historical significance and stunning natural beauty as well as for its archaeological richness, a fact not lost on Steven Spielberg, who filmed the last scenes of *Indiana Jones and the Last Crusade* here. Its natural rock fortifications served to protect not only the Nabatean and Roman armies, but T. E. Lawrence and members of his Arab militia, who briefly lived in some of the 500 caves during their guerrilla campaign in World War I.

The ruins are located in a geological formation known as the Dead Sea Rift, which was formed 30 million years ago when the Arabian plate separated from Africa, forming the Red Sea, the Gulf of Aqaba and the Jordan Valley. While the sandstone rock of the area is primarily reddish, rose and white, it is also streaked with ribbons of yellow and purple. Visitors must approach the site on horseback down the narrow Petra Valley, a twisting gorge with sheer rock cliffs that opens dramatically onto one of the most spectacular Nabatean tombs, the Khazneh. In 1911, a German scholar, Gustaf Dalman, called the tomb "the most perfect two-storied facade which has been preserved in the East from antiquity until now."

The area is mentioned several times in the Bible. The valley itself is often called the Valley of Moses, and nearby is a rise that both Muslims and Christians believe to be the Biblical Mount Hor, the burial place of Aaron.

The excavation of the Byzantine church, which is located on a ridge above the Roman road, is being done by the American Center for Oriental Research with a $500,000 grant from the United States Agency for International Development. The church was discovered in 1990 by Dr. Kenneth W. Russell of the American center. Two mosaics, each 60 square meters (about 72 square yards), have been found, each running down an aisle. Much of the work of the last year and a half has focused on the restoration and cleaning of the mosaics. Dr. Russell died of tick typhus in 1992, just as the work was beginning.

The mosaics, considered to be some of the finest uncovered in Jordan, depict a variety of figures, including numerous birds, pairs of elephants, camels, wild boars and giraffes. There are also a flute player, a fisherman and a camel driver, as well as figures that represent each of the four seasons.

The Nabateans rose to prominence in Jordan during the later stages of the Iron Age, roughly corresponding to the period covered by much of the Hebrew Bible. They built elaborate aqueducts, many of which are still visible, and a system of clay pipes to bring water in from distant springs. The Nabatean script evolved into modern written Arabic. In the Augustan age, the kingdom was responsible for 25 percent of the gross economic output of Rome.

Petra was the capital of the Nabatean kingdom. Because of its strategic location—along the incense, silk and spice route that wound its way from China, India and southern Arabia to Rome—many Nabateans became wealthy, and Petra became a center for artists and scholars.

When the kingdom was annexed by Emperor Trajan of Rome in A.D. 106, it became one of the empire's most important trading centers. In the fourth century, it continued to be an important political outpost for the Byzantine Empire, becoming the seat of a bishopric. But the city soon began to fall into decline. Petra sank into obscurity after a shift in trade routes that was followed by two powerful earthquakes, one in A.D. 363 and a second in 551.

"After the earthquake in 363 A.D., you see that many structures were not rebuilt," Dr. Fiema said. "They apparently just did not have the economic resources available in the past. If you look at the shops along the colonnaded street, you see that many shopkeepers simply rebuilt their structures in front of those that had collapsed in the earthquake, rather than bothering to clear away the rubble. It is a sign that the wealth and order of the city was beginning to decline."

Many of the buildings, including the sixth-century church under excavation, appear to have burned as well as collapsed.

The desolation that fell over the city helped preserve it. In 1812, a Swiss explorer, Johann Ludwig Burckhardt, disguised himself as an Arab traveler and slipped into the valley, then recorded its location.

But little was done to excavate the remains. German archaeologists worked at the site at the beginning of the century. And a team from the University of Utah, under the direction of Dr. Philip C. Hammond, excavated a theater and continues to work on the imposing Temple of the Winged Lions. A British team led by Peter Parr rebuilt an arched gate in 1958 and 1959. The Qsar al-Bint temple, perhaps the most important building in Petra, has been under excavation by a Jordanian team led by Dr. Fawzi Zayadine for the last decade. A team from the University of Basel in Switzerland has also spent several seasons excavating.

Petra's potential has a hypnotic grip on those who bear the searing heat and dust to uncover the remains of the city.

"Just a hundred meters from us is a huge pile of stones and a granite column protruding from the ground," said Dr. Fiema, pointing to the side of a nearby hill. "Granite is not available in Jordan. It must have come from Egypt. I often look and wonder what structure lies beneath the ground.

"Is it a royal palace? A temple? Wherever you walk in Petra, you are faced with such puzzles."

—CHRIS HEDGES, January 1994

# Ancient Graves of Armed Women Hint at Amazons

THE ANCIENT GREEKS COULD certainly tell a good yarn. Cultivating a kernel of fact, or less, they could bring forth a feast of a story to nourish imaginations down through the ages. One such tale was about a society of fierce warrior women—the Amazons.

In the account by the historian Herodotus in the fifth century B.C., Greek soldiers in the Black Sea region found themselves in combat against an army of women. Although the Greeks won, their foe made a lasting impression. Here were women who did not confine themselves, as Greek women did, to cooking and weaving and other domestic roles. They lived to fight and were required to kill an enemy before marrying. They even cut off their right breasts, the better to shoot with bows and arrows.

Or so the story went. Herodotus conceded that he had never seen an Amazon; his tale was based on hearsay. But archaeologists excavating graves in the Eurasian steppes are now finding evidence that there may be something to the Amazon legend after all.

Over the last four years, American and Russian archaeologists have examined 44 mounds, or kurgans, near the town of Pokrovka in Kazakhstan at the Russian border, where ancient nomad cultures buried their dead. From the grave goods and other evidence, the burials appeared to be associated first with the Sauromatians and then the early Sarmatians, Indo-European–speaking herders who lived on the steppes in the sixth to fourth centuries B.C. and fourth to second centuries B.C., respectively. But the most striking discovery at Pokrovka has been the skeletons of women buried with swords and daggers. One young woman, bowlegged from riding horseback, wore around her neck an amulet in the form of a leather pouch containing a bronze arrowhead. At her right side was an iron dagger; at her left, a quiver holding more than 40 arrows tipped with bronze.

"These women were warriors of some sort," said Dr. Jeannine Davis-Kimball, a leader of the excavations. "They were not necessarily fighting battles all the time, like a Genghis Khan, but protecting their herds and grazing territory when they had to. If they had been fighting all the time, more of the skeletons would show signs of violent deaths."

In that case, the Sauromatian-Sarmatian women probably did not quite fit the larger-than-life Amazon image of women who seemed to prefer making war to making love. Also, the women at Pokrovka lived more than 1,000 miles east of the Amazons the Greeks supposedly encountered. So Dr. Davis-Kimball is not jumping to any conclusions that these women were indeed the Amazons of legend, only suggesting that they could be contemporaries of the Amazons or that their lives, and those of similar nomadic women who could ride and wield a sword or dagger in combat, may have inspired the legend.

In the earlier Sauromatian graves, the skeletons revealed one suggestive Amazonian attribute. The men and women, at an average of five feet, 10 inches and five feet, six inches, respectively, were taller and more robust than normal people at that time.

Of more importance, the new discoveries are forcing anthropologists and historians to reconsider the status and role of women in the Eurasian nomad societies of the first millennium B.C. The research, she said, showed that women seemed to have more wealth, power and status in these cultures than anyone had thought. And certain women, perhaps the elite of the tribe, appeared to be trained from an early age to be warriors on horseback.

Dr. Davis-Kimball, an archaeologist at the Center for the Study of Eurasian Nomads in Berkeley, California, described the Pokrovka research in an article in *Archaeology* magazine and in a more scholarly report published in the *Journal of Indo-European Studies.* She and other specialists in central Asian archaeology discussed the interpretations in interviews.

In her analysis, Dr. Davis-Kimball said burials at Pokrovka and other sites seemed to reveal three categories for women of the culture. Graves with luxury goods, including beads, colored glass and gilted earrings, suggested that the "most frequently found status among females," she said, "is that of femininity and the hearth." The women in a few graves might have been priestesses; they were buried with stone altars, bronze mirrors used in healing and other cultic materials. Finally, there were the warrior women.

Dr. Nicola DiCosmo, a historian of central Asia at Harvard University, said that other archaeological findings in the steppes from Russia to Mongolia seemed to indicate that Dr. Davis-Kimball "is on to something." The findings, he said, showed that "women in early nomadic societies could have had a higher profile in their cultures than women in sedentary societies at the same time."

Dr. Elizabeth J. W. Barber, an archaeologist at Occidental College, said the research represented a significant change in the most rudimentary level of archaeological interpretations. Until recently, she said, "most people assumed that if a grave had weapons, the skeleton was a man—now they can't be so sure."

Some Russian archaeologists who had made similar discoveries at other sites have argued that the weapons found with female burials had nothing to do with a person's life but were placed there for protection in the afterlife. But Dr. Davis-Kimball points to the bowed leg bones and amulets seeming to denote prowess in the hunt and battle to dispute such an explanation.

"Probably, we've been carried away with the macho image of the nomad," said Dr. Claudia Chang, an anthropologist at Sweet Briar College in Virginia, who conducts excavations in Kazakhstan. She noted accumulating research indicating that women in these ancient cultures sometimes "had an active role in warfare and in the political structure." But she cautioned against "ascribing more to the women of these cultures than actually existed."

In the article in *Archaeology* magazine, Dr. Davis-Kimball said the excavations showed that the nomad women seemed "to have controlled much of the wealth, performed rituals for their families and clan, rode horseback and possibly hunted saiga, a steppe antelope, and other small game." In times of crisis, she wrote, the women "took to their saddles, bows and arrows ready, to defend their animals, pastures and clan."

And, perhaps, to astonish Greek soldiers and inspire enduring yarns.

—JOHN NOBLE WILFORD, February 1997

# New Analysis of the Parthenon's Frieze Finds It Depicts a Horrifying Legend

STANDING IN COLUMNED SPLENDOR atop the Acropolis of Athens, the Parthenon is the paragon of classical architecture and has long been a shrine of Western civilization. The Greeks built this temple to the goddess Athena in the fifth century B.C., in the golden age of Pericles. Marble statues portrayed the mortal and immortal greats, and at the center was a huge cult statue of Athena.

Set high above, on all sides in the shadow of the exterior colonnade, was a 524-foot-long frieze in low relief depicting various stages of what appeared to be a single solemn ceremony. A cavalcade of mounted soldiers was followed by people bringing animals to sacrifice and bearing offerings by musicians, maidens and elders. They approached a central scene above the east entrance, where among other figures, a man and child held a large cloth.

The interpretation of these scenes is now facing a serious challenge for the first time in two centuries.

The received wisdom on the subject stems from a report by two British travelers in 1787. The artist James Stuart and the architect Nicholas Revett, members of the Society of Dilettanti in London, returned from a visit to the Acropolis with drawings, descriptions and their interpretation that the frieze represented the Panathenaic festival, which was held every four years to commemorate the birth of Athena. The piece of cloth was presumably a new robe, or peplos, being presented for draping over the Athena statue as the culminating ritual of the festival.

Now the discovery of fragments of a lost play by Euripides, found on papyrus in the wrapping of an Egyptian mummy, and the diligent research of an American archaeologist have produced a much different explanation. The scenes of the frieze do not depict a fifth-century procession, according to the new thesis, but instead evoke the Athenian founding myth of a king's precious sacrifice to save his city from defeat.

Such an interpretation may be more satisfying to scholars and more revealing of early Greek culture and mythology—but it may also become controversial and eventually even disillusioning. To think that this iconic structure of grace and just proportion could turn out to have been dedicated to the glorification of a practice as primitive, cruel and irrational as the sacrifice of children! And worse, that it dated from the time the Greeks were boldly experimenting with democracy and rationalism, from that age whose creative spirit the Renaissance sought to emulate.

In a close reading of the legend of Erechtheus, the heroic king of early Athens, Dr. Joan Breton Connelly, an associate professor of fine arts at New York University, came to realize that the peplos scene in the frieze could represent the sacrifice of the young daughters of the king. This was the price required of him by the oracle of Delphi if Athens was to be saved from its besieging foes.

The newly discovered Euripidean text, she said, shows that the story of Erechtheus and virgin sacrifices for the good of the city were themes resonating among Athenians at the time the Parthenon was being built in the heady years following their defeat of the Persians at Marathon.

Viewed in this light, the five individuals in the peplos scene could be the mythic royal family: Erechtheus, his wife, Praxithea, and their three daughters. The three girls may well be preparing for death. The youngest girl's funerary dress is being unfolded; she will go first. The oldest daughter, second from the left, is in the process of handing down a stool to her mother. The daughter at the far left faces to the front, with a garment still folded and carried upon the stool on her head.

Dr. Connelly said this new interpretation "has far-reaching implications for our understanding of the role of women in Greek myth and culture." Greek writings of the time were making much of the sentiment attributed to Praxithea that just as boys go to war, girls go to sacrifice—both for the good of the polis, the city-state. Another famous example was Agamemnon's sacrifice of his daughter Iphigenia, which enabled the Greek fleet to set sail for war against the Trojans.

In a paper prepared for publication in the *American Journal of Archaeology,* Dr. Connelly concluded that the new explanation for the frieze "encourages us to reevaluate our current understanding of the Panathenaic festival itself, an event which may have been more than just the celebration of Athena's birthday."

Homer wrote in the *Iliad* of funeral games established in memory of deceased heroes. So, too, the festivities and games for Athena's birthday might have served as a commemoration as well of the Athenian founding myth of Erechtheus and his daughters.

"I think she may be proved right," said Dr. Homer A. Thompson, an emeritus professor at the Institute for Advanced Study in Princeton, New Jersey, who at the age of 88 is the grand old man of classical Greek archaeology.

Dr. William St. Clair, a fellow of All Souls College at Oxford University and author of *Lord Elgin and the Marbles,* called Dr. Connelly's thesis nothing short of revolutionary, and not unreasonable.

Writing in the *Times Literary Supplement* of London recently, Dr. St. Clair said, "It is surprising in retrospect that nobody familiar with the conventions of Greek temple decoration" had thought of the new interpretation before.

He was alluding to one problem with the usual explanation for the frieze that has long troubled scholars. Why would the Parthenon frieze record a contemporary event? The convention in Greek architecture was to decorate sacred buildings with scenes from myth; to do otherwise, as a scholar once wrote, "verged on profanation."

Other elements in the frieze also raised suspicions. If this was indeed a Panathenaic procession, where was the wheeled ship that should have been drawn to the temple bearing the sacred peplos of Athena? Where were the maiden basket bearers, the armed foot soldiers or the women water carriers? There were male water carriers, who should not have been there, and also chariots, which were not used in combat by classical armies. With no satisfactory answers at hand, most scholars have chosen to ignore such questions and cling to standard explanations, but not Dr. Connelly.

Although she has yet to publish details of her interpretation, she has described it in lectures, particularly in England, where she just completed a year doing research at Oxford. She wanted to be near many sections of the frieze, which were removed from the Parthenon by Thomas Bruce, Earl of Elgin (Lord Elgin), in the early nineteenth century and are now at the British Museum in London.

More than four fifths of the frieze survives, mainly in museums in Paris, Rome and Athens as well as London. Wars and air pollution have taken their toll on the rest of the Parthenon, with explosives toppling and destroying statues and columns and corrosive air eating away at the marble and limestone. The temple is undergoing extensive restoration under the direction of Manolis Korres, a Greek architect.

In an interview by telephone from Cyprus, where she is directing archaeological excavations unrelated to her Parthenon research, Dr. Connelly said she began to question accepted ideas about the frieze years earlier while studying Greek priestesses. Icons in the frieze, she said, did not appear to be those usually associated with the priesthood.

At the same time, she happened to be reading the rediscovered lines from the Euripides tragedy about Erechtheus, seeking insights about priestesses. The play was written about 423 B.C., a decade after the completion of the Parthenon. But only 125 lines survived, until French technicians in 1962 recovered those papyrus fragments from a mummy at the Louvre in Paris. They managed to peel them away from the mummy wrappings without destroying the ink writing. Later, Dr. Colin Austin, a scholar at Cambridge University in England, determined that the writing was 120 lines of the lost Euripidean play.

Reading those lines, Dr. Connelly recognized that the woman in the peplos scene was no ordinary priestess, but probably the queen of the myth, Praxithea. In the myth, she has not only lost her daughters but her husband Erechtheus, who perished in his triumphant battle against Eumolpos, son of Poseidon. In the play's recovered lines, Athena appears to Praxithea with instructions for the burials of the king and the daughters on the Acropolis and for remembering them "with annual sacrifices and bull-slaying slaughters" and "with holy choruses of maidens."

As for Praxithea, Athena rewarded her for restoring "the foundations of the city" by designating her a "priestess to make burnt sacrifice at my altar on behalf of the city."

Dr. Connelly said she felt her ideas were being well received, though Dr. St. Clair of Oxford was not so sure. "The boldness of Joan Connelly's suggestion has left other scholars silent or dismissive," he said. "They cannot bring themselves to accept it, yet so far at least, they appear reluctant to confront the arguments directly."

Despite the excitement of advancing revolutionary ideas about one of the world's most famous buildings, Dr. Connelly hopes someday to resume her interrupted research on Greek priestesses. "All this has told me more about the Parthenon than about priestesses," she said, not really as a lament.

—JOHN NOBLE WILFORD, July 1995

# Archaeologists Unearth Treasure Buried by the Cold War

POMPEII, THE DECREPIT, graffiti-scarred heap at the foot of Mount Vesuvius, is probably the world's most famous archaeological village. Frozen forever by a volcanic eruption 2,000 years ago, it has long served as ancient history's Disneyland.

But Pompeii provides a lucky snapshot in the life of a stately Roman town. Chersonesos, covering the edge of this Crimean city, presents grander possibilities: 2,500 years ago, this was the Greek world's most northern colony. It was here, along the Black Sea coast, that "civilized" colonists first encountered Scythians, the nomadic "barbarians" from central Asia.

Spread across hundreds of urban and rural acres, Chersonesos (it is from the Greek word for "peninsula" and pronounced CHER-so-NEE-sus) is the most important classical site on the Black Sea. Scythian tombs and Roman fortresses lie under the rolling hills nearby. Chersonesos was the Byzantine world's largest trading outpost on the Black Sea until 1399, when it was sacked by the Mongols, and this was the port through which Christianity entered Russia. That's not all. Here, from the earthen redoubts of Balaclava, one can gaze through the mists upon Tennyson's "valley of death," where in October 1854 the British cavalry was savaged by the Russian army in an infamous act of military folly during the Crimean War. Since then, the battle has been called "The Charge of the Light Brigade."

"The historical significance of this area and this site is really hard to exaggerate," said Dr. Joseph Coleman Carter, a professor of classics at the University of Texas and the director of its Institute of Classical Archaeology. Dr. Carter has spent much of his career explaining the migratory and rural land-use patterns of the ancient Greeks.

But until 1992, he was never able to range as widely as the Greeks had themselves. Chersonesos was closed to foreigners because it was near the headquarters of the Soviet Black Sea Fleet and its nuclear submarine base in Sevastopol. Although the more urban part of the site was famous even during the time of the czars, when much money was spent to excavate here, photographs and maps were almost impossible to get. When the city was opened in 1992, Dr. Carter was the first Western archaeologist invited to work here by his Ukrainian colleagues. One of his goals is to "make the Greek world one again" by linking its ancient colonies.

Working with Ukrainian archaeologists to excavate the sites, Mr. Carter has a grand vision of turning Chersonesos into one of the world's great archaeological preserves, a park filled with treasures like ancient farmhouses and forts and granaries that would explain how civilization evolved here. From his base on the Crimean coast, one of the most attractive stretches of vacation land in the former Soviet Union, he dreams of an archaeological paradise for tourists.

"There is no place on earth like Chersonesos," he said during a recent visit to the site. "It is the Russian Pompeii. Greeks, Romans and Byzantines all had their day. Every great epoch built its way of life on this soil. There are forts, mints and farmhouses. If we could restore what is here and present that to people, it would be remarkable."

It would also require millions of dollars and the resolution of several political disputes, most importantly one between those who wish to preserve the site and the Ukrainian Orthodox Church, which says it owns most of this priceless land. Church officials object to its pagan monuments on religious grounds.

The battle for Chersonesos has already been joined. There are 30,000 acres on the Crimean Peninsula. Of those, 1,000 are protected by law from development. In 1996, Chersonesos became the only national preserve in Ukraine, which means that cultural authorities regulate its development in this extremely popular coastal region. But in late summer, 1997, church officials in a helicopter entered the site, deposited a roof on a baptistery and asserted that the whole land was theirs. Church officials are not shy about stating that they want pagan monuments—and there are hardly any other kind here—destroyed. They have even referred to the director of the preserve as a devil.

So far, the Culture Ministry has sided strongly with the archaeologists. So have many others in Ukraine and throughout the world. But it would cost millions of dollars to produce a park on the scale envisioned by those involved in the preservation here. The church, which has been a staunch opponent, is powerful, and so is the temptation to bring in revenues through vast and ruinous private development. It is partly for that reason that the ancient city and territory have for the past two years appeared on the World Monument Fund's watch list of the hundred most-endangered cultural sites.

The World Bank and other agencies interested in preserving culturally significant sites have begun to consider financing the project here. The idea of the park has already found the support of the Samuel H. Kress Foundation, which has a long history of supporting archaeological preservation.

For Dr. Carter, this work—even its scale—is nothing new. Before turning his attention to the northern boundaries of Greek civilization, he spent more than two decades exploring and documenting the rural archaeology of a vast Greek colony in Metaponto, a town in southern Italy. To uncover all that is here would take at least that long, he said.

Because Chersonesos is considered the best-preserved Greek colonial territory, no site in this part of the world could make a more vivid contribution to understanding the rural roots of ancient civilizations. Dr. Carter and his Ukrainian and Russian colleagues hope to develop a series of sites that would represent the life of the territory from its earliest inhabitants—the oldest artifact found dates from the sixth century B.C.—through the Greek, Roman and Byzantine eras. Each aspect of life here would be illustrated by a different monument or rural site.

The farmlands are in many ways what interest the archaeologists most. And by linking them to those in southern Italy, Dr. Carter and other archaeologists hope to be able to learn more about the history of the colonies and of land distribution.

"From the eighth century B.C., the Greeks were looking for land beyond Athens that could sustain them," Dr. Carter said. "By then the city was overpopulated and expansion was essential. People assume mistakenly that those who lived in the country were economically and socially inferior to those in the Greek cities. It wasn't true at all."

Ancient Greece was the definitively urban society. The rich cultural heritage in Metaponto demonstrated convincingly for the first time, however, that rural life was just as important to the Greeks. There was almost no such thing as an intact rural heritage near Athens. But in Metaponto, and in Chersonesos, farm divisions have been so perfectly preserved over thousands of years that ancient fence lines still appear clearly in aerial photographs of the region. And this was the breadbasket of the ancient Greek world. The soil around Athens was notoriously rocky and poor; here it is rich and fertile. By the third century B.C., it had become the major source of grain for Athens.

"The division of the countryside was really the basic division of Greek society," said Galina Nikolaenko, deputy director of the archaeological museum here and the director of the dig in the agricultural territory. "That was really the core of democracy. And this we never really realized until now."

Many farmhouses were the sites of family cults. There might have been a single piece of sculpture in the house to be used in rural sacrificial rites. A mint, which may have served the city as well as the rural suburbs, has been discovered dating from the fourth century B.C., a time when the city probably had 10,000 to 20,000 residents.

The site here overlooks the valley that is best entry into the area, and it was always a place of military significance.

The Greeks used it as a fortress against the Scythians. Then the Romans built a fort here to repel the Goths and, more than 1,000 years later, so did the Byzantines against the Huns. (It was also the place where the light brigade was so famously destroyed.) This summer, the team of archaeologists, some from Texas and the rest Russian and Ukrainian, found evidence of an extensive village dating to the fourth, fifth and sixth centuries B.C. on a hilltop not far from where the Battle of the Light Brigade was fought.

When digging, the team often had to sift through a mixture of artifacts that ranged from 2,500-year-old items to those from this century, like leather holsters and bullets from the World War II siege here that killed 150,000 Russian and German soldiers.

"If we could only do the research we need to do," said Leonid Marchenko, the energetic director of the local museum, "we will find a history we cannot yet even imagine. It's been here for thousands of years. It's all in one place. We just need to let it out."

—MICHAEL SPECTER, November 1997

# Children's Cemetery a Clue to Malaria as Rome Declined

DIGGING AMONG THE RUINS of a Roman villa, archaeologists have made a macabre discovery about disease and death in the fifth century A.D., and perhaps even about the reasons for Attila the Hun's decision to leave his invasion of Italy unfinished and for the decline and fall of the Roman Empire.

The discovery is a cemetery for infants, excavated over the last two years by an international team led by Dr. David Soren, a classical archaeologist at the University of Arizona in Tucson. The cemetery overlooks the Tiber River near the town of Lugnano in Teverina, 70 miles north of Rome.

With 49 skeletons already uncovered and excavations not complete, this is the largest ancient cemetery for infants ever found in Italy, Dr. Soren and other members of his team said in interviews. The tiny skeletons bear the first apparent physical evidence of the epidemics known from literature to have plagued imperial Rome, especially in its latter centuries. The new evidence, some direct but most circumstantial, points to malaria as the cause. That is not surprising because the marshes around Rome in earlier times were breeding grounds for mosquitoes, the source of the summer "vapors" blamed for leaving people weak or dying with fever.

This association has been so strong that the word "malaria" comes from the Italian for "bad air." The threat of the "Roman fever" has persisted almost to the present, as Henry James knew in writing his 1878 story *Daisy Miller.* When the guileless young American took her fateful stroll by moonlight in the Colosseum, she ignored warnings of the "villanous miasma" and died a few days later of "a terrible case of the fever."

The hasty multiple burials in the infant cemetery tell of an epidemic's swift toll long ago. The presence of decapitated puppy skeletons, a raven's claw and other examples of pagan ritual seems to reflect the desperation of a people who, though by this time officially Christian, revived witchcraft and superstitious offerings in their moment of extreme stress.

"Just about everything about this site is odd and interesting," Dr. Soren said. "The preservation of the skeletons is remarkable, and there are all those puppies. There's nothing of Roman gods, not anything Christian in the place, only what might be called village witchcraft."

But the Christian influence must have been established by then, or people would not have even thought to have a cemetery where newborn children were given proper burials. Since Christians baptized infants and considered them significant humans at least from birth, they could not merely discard the bodies of dead infants or bury them unceremoniously within houses, as had been the earlier Roman practice. Indeed, the Romans often put newborn and sickly infants out to die; female babies, in part because of their later dowry obligations, were always at risk.

The discovery, made by Dr. Soren's team in conjunction with the Antiquities Service of Umbria, a government agency in Italy, was an outgrowth of the team's earlier investigations of the ruins at Villa Poggio Gramignano, described in *Archaeology* magazine. Built around the time of Christ and in use until the third century, the villa was unusual architecturally in that it had a pyramid-shaped ceiling over its main colonnaded reception hall. But the shifting hillside bedrock undermined the walls, causing the villa's collapse into ruins.

The uniformity of the pottery and other artifacts among the graves, Dr. Soren said, suggests that all the burials occurred over a brief period around the year 450. Most of the infants were interred in earthen jars. At the lower levels, most graves contained a single skeleton; none held more than two. At higher levels, there were mass graves, each with five or six infants.

Such a pattern indicates that the death rate in the community might have been normal at the time of the first burials, but it suddenly escalated, as from an epidemic. No adult graves have been found at the site.

An examination of the skeletons by medical scientists reveals that some of those buried were premature infants. Others were neonates, no more than a month old, and others were up to five or six months old. The skeleton of one child of two or three years was found. The older children were generally buried in more elaborate graves, but the others were often interred amid refuse from the abandoned villa, further evidence, Dr. Soren said, of the Roman belief that newborn infants were not "worthwhile family members and should not be lamented much if they died."

Although the bones of the youngest infants were too fragile for detailed analysis, Dr. Jose Ribeiro, a medical entomologist at Arizona, noted that five of the older skeletons exhibited a condition known as porotic hyperostosis, spongy and pitted bone, on the external surface of the cranium.

Such damage could be caused by severe anemia brought on by malnutrition. But at the site archaeologists found an abundance of animal bones, particularly pig, so they tended to discount famine as a major factor in the deaths. Porotic hyperostosis could also be a body's response to an infectious disease like malaria.

Dr. Ribeiro said the falciparum parasite, which is spread by the anopheles mosquito and causes malaria, is known to have a grave effect on pregnant women, causing aborted and premature fetuses, stillbirths and infant deaths. A malaria epidemic would have begun slowly and caused multiple deaths over a period of weeks or a month. The Tiber basin, with its marshes of stagnant water in those times, would have easily supported the disease-carrying mosquitoes.

From the evidence so far, Dr. Soren said, "the likely conclusion, based on the pattern of the burials, availability of food and the bone analysis, is that malaria was the agent of death."

The archaeologists cautioned that this was only a preliminary hypothesis, requiring more excavations and research, but they have been encouraged by a close investigation of historical and literary records.

"We know from written records," said Dr. Frank Romer, a classics professor at Arizona, "that there were a series of strange pestilences in the middle of the fifth century, a couple of them quite clearly affecting this area near Rome."

These epidemics were no respectors of conquerors. Soon after Alaric I, the Visigoth king, sacked Rome in 410, he suddenly died of an illness, which some historians suspect was malaria. When Attila led the Huns into Italy in 452, he inexplicably turned back short of Rome, after a meeting with Pope Leo I. Perhaps the pope had told him an army from Byzantium was on its way to reinforce Rome's defenses. Or perhaps, Dr. Romer said, the pope had frightened Attila off with accounts of appalling sickness throughout the land.

In his research, Dr. Romer also came across a law passed in January 451, the time of the cemetery, that addressed the problem of selling children, a problem brought on by a health crisis throughout Italy. Parents were compelled to sell children, the law noted, in the hope that their chances of survival might be better elsewhere, perhaps overseas or in supposedly healthier areas in the north.

In 467, a wealthy nobleman from Gaul, Sidonius Apollinaris, wrote from Italy of the pestilence spreading through the country. He became ill himself, and the symptoms he described were much like those of malaria. He wrote that his body "had been suffocated by breathing the air which is imbibed in poisoned gasps and which alternates sweats and chills."

Until now, Dr. Romer said, historians have not fully appreciated the extent of the human suffering described by the nobleman. Various plagues had afflicted the Roman Empire since the second century and contributed to its decline in strength. By the fifth century, the seat of power had shifted to Byzantium, leaving Italy a mere province. Not until 1929 did the Italians drain the marshes around Rome and all but eliminate the scourge that may have shaped history and left all those tiny skeletons in the villa ruins on a Tiber hillside.

—JOHN NOBLE WILFORD, July 1994

# Ice Cap Shows Ancient Mines Polluted the Globe

SAMPLES EXTRACTED FROM Greenland's two-mile-deep ice cap have yielded evidence that ancient Carthaginian and Roman silver miners working in southern Spain fouled the global atmosphere with lead for some 900 years.

The Greenland ice cap accumulates snow year after year, and substances from the atmosphere are entrapped in the permanent ice. From 1990 to 1992, a drill operated by the European Greenland Ice-Core Project recovered a cylindrical ice sample 9,938 feet long, pieces of which were distributed to participating laboratories. The ages of successive layers of the ice cap have been accurately determined, so the chemical makeup of the atmosphere at any given time in the past 9,000 years can be estimated by analyzing the corresponding part of the core sample.

Using exquisitely sensitive techniques to measure four different isotopes of lead in the Greenland ice, scientists in Australia and France determined that most of the man-made lead pollution of the atmosphere in ancient times had come from the Spanish provinces of Huelva, Seville, Almeria and Murcia. Isotopic analysis clearly pointed to the rich silver-mining and smelting district of R o Tinto near the modern city of Nerva as the main polluter.

The results of this study were reported in the journal *Environmental Science & Technology* by Dr. Kevin J. R. Rosman of Curtin University in Perth, Australia, and his colleagues there and at the Laboratory of Glaciology and Geophysics of the Environment in Grenoble, France.

One of the problems in their analyses, the authors wrote, was the very low concentrations of lead remaining in ice dating from ancient times—

only about one hundredth the lead level found in Greenland ice deposited in the last 30 years. But the investigators used mass-spectrometric techniques that permitted them to sort out isotopic lead composition at lead levels of only about one part per trillion.

Dr. Rosman focused on the ratio of two stable isotopes, or forms, of lead: lead-206 and lead-207. His group found that the ratio of lead-206 to lead-207 in 8,000-year-old ice was 1.201. That was taken as the natural ratio that existed before people began smelting ores. But between 600 B.C. and A.D. 300, the scientists found, the ratio of lead-206 to lead-207 fell to 1.183. They called that "unequivocal evidence of early large-scale atmospheric pollution by this toxic metal."

All ore bodies containing lead have their own isotopic signatures, and the R o Tinto lead ratio is 1.164. Calculations by the Australian-French collaboration based on their ice-core analysis showed that during the period 366 B.C. to at least A.D. 36, a period when the Roman Empire was at its peak, 70 percent of the global atmospheric lead pollution came from the Roman-operated R o Tinto mines in what is now southwestern Spain.

The R o Tinto mining region is known to archaeologists as one of the richest sources of silver in the ancient world. Some 6.6 million tons of slag were left by Roman smelting operations there.

The global demand for silver increased dramatically after coinage was introduced in Greece around 650 B.C. But silver was only one of the treasures extracted from its ore. The sulfide ore smelted by the Romans also yielded an enormous harvest of lead.

Because it is easily shaped, melted and molded, lead was widely used by the Romans for plumbing, stapling masonry together, casting statues and manufacturing many kinds of utensils. All these uses presumably contributed to the chronic poisoning of Rome's people.

Adding to the toxic hazard, Romans used lead vessels to boil and concentrate fruit juices and preserves. Fruits contain acetic acid, which reacts with metallic lead to form lead acetate, a compound once known as "sugar of lead." Lead acetate adds a pleasant sweet taste to food but causes lead poisoning—an ailment that is often fatal and, even in mild cases, causes debilitation and loss of cognitive ability.

Judging from the Greenland ice core, the smelting of lead-bearing ore declined sharply after the fall of the Roman Empire, but gradually increased during the Renaissance. By 1523, the last year for which Dr. Rosman's group conducted its Greenland ice analysis, atmospheric lead pollution had reached nearly the same level recorded for the year 79 B.C., at the peak of Roman mining pollution.

—MALCOLM W. BROWNE, December 1997

# 4

# BIBLICAL ARCHAEOLOGY

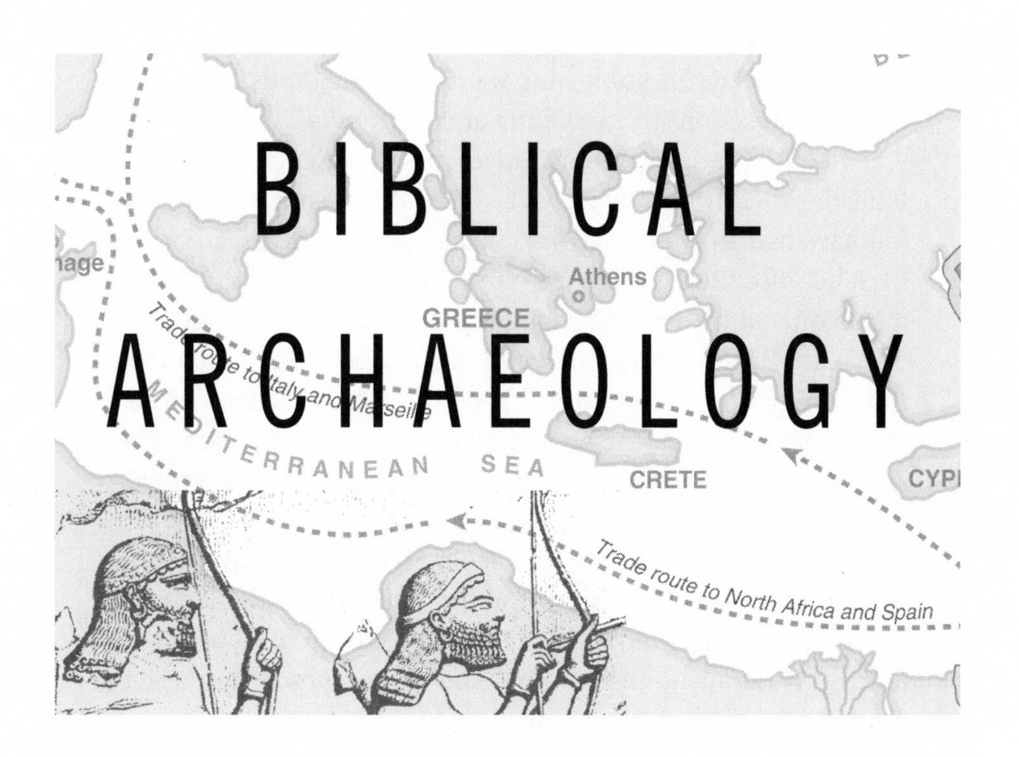

The archaeology of the biblical world has a special extra dimension: It relates to religion as well as history. Not surprisingly it often seems to be engulfed in furious controversy of one sort or another, whether in criticisms of the inordinate delays in publishing the Dead Sea Scrolls or in ultra-religious objections to excavating ancient tombs.

Recent finds have included new evidence about the Philistines, early inhabitants who gave their name to Palestine, but have had an enduringly bad image, perhaps deserved or perhaps because their opponents wrote the history books. No written records of the Philistine language have yet come to light.

The Philistines first come into history because of their invasion of Egypt, as recorded in Egyptian records. When defeated there, they settled in the coastal plain from Joppa to Gaza. They soon came into conflict with the Israelites and were eventually defeated by King David.

The detective work of archaeologists has recently been rewarded with the discovery of the first mention of the House of David, outside the Bible, at the ancient city of Tel Dan in the form of an Aramaic inscription. And the origins of mysterious tunnels in Jerusalem have come to light.

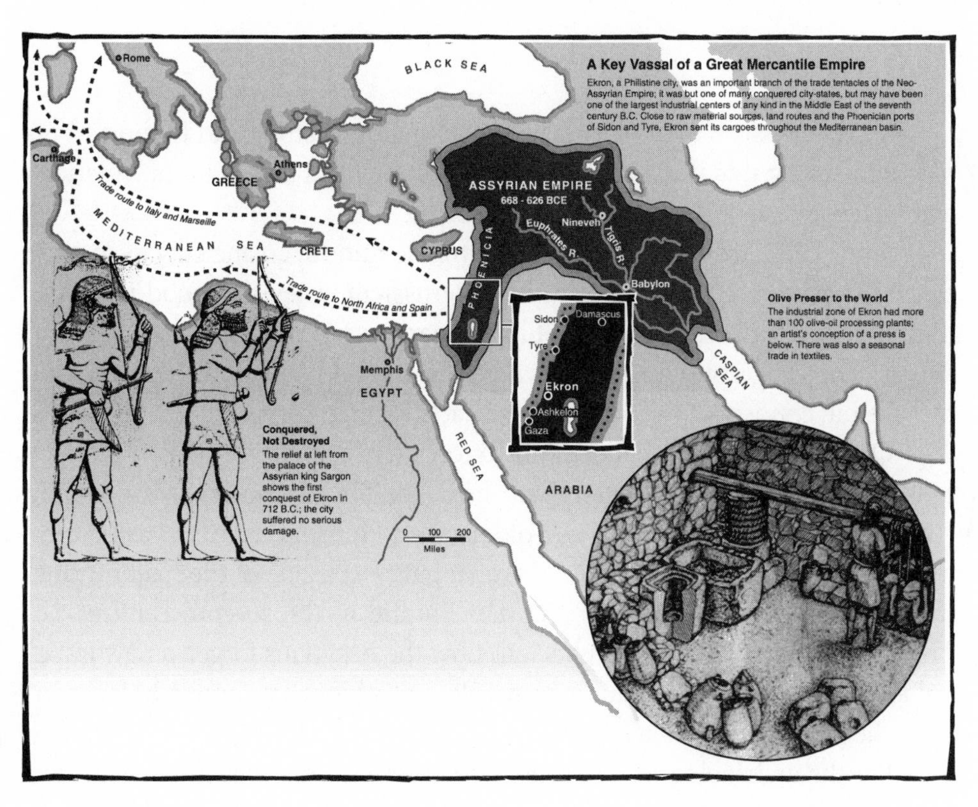


The New York Times

# Inscription at Philistine City Shows: This Is the Right Place

ARCHAEOLOGISTS DREAM OF turning over a temple stone and finding an inscription saying this is the place you are looking for. For a team of American and Israeli archaeologists, the dream came true.

Since 1983, they have been excavating the ruins of an ancient city at a site called Tel Miqne, 20 miles southwest of Jerusalem. They had good reason to think this was the Philistine city of Ekron, mentioned in the Bible and Assyrian annals. The geography was right: where the coastal plane of ancient Philistia met the hill country of Judah. All the artifacts seemed recognizably Philistine.

On the assumption that this was Ekron, archaeologists and other scholars examining the decorated pottery and evidence for advanced town planning concluded that contrary to the age-old slander, Philistine culture was no oxymoron. They could also see that this must have been one of the major industrial cities of the far-flung Neo-Assyrian Empire in the seventh century B.C. That gave them important insights into how the Assyrians forged a new imperial ideology based on mercantile principles, creating what some scholars consider the first "world market."

But the archaeologists could not be absolutely sure that this was indeed Ekron until Dr. Seymour Gitin, director of the Albright Institute of Archaeological Research in Jerusalem, turned over a large block of stone found near the entrance to a colonnaded building at Tel Miqne. His expectations were low because nothing with writing had been found there yet.

"I had been getting tired of this," Dr. Gitin said in a telephone interview from the dig site. "Nothing ever showed up."

When the caked dirt was cleared away, though, he let out an unscholarly exclamation: "Oh, my God!" He saw a five-line inscription written in Phoenician script, and some of the 69 letters spelled out the name Ekron and the names of two of the city's known kings, Achish and his father Padi. The inscription recorded that Achish had built a temple here dedicated to a goddess.

"We always felt this was Ekron, but to find the inscription makes the identification one hundred percent," Dr. Gitin said. "This you don't find very often in archaeology."

In fact, he said, this is the first time the name of a biblical city and a list of its kings has ever been found on a site where its historical context is clear. No other such monumental inscription has been found in Israel from the biblical period. Other scholars agreed.

"It's a very exciting find," said Dr. Gary A. Rendsburg, a professor of Near East studies at Cornell University.

For one thing, the inscription could give scholars the first strong evidence of the language of the Philistines. They were descendants of the enigmatic Sea People, originally from the Aegean Sea region, who arrived in large numbers on the coast of Canaan soon after 1200 B.C. Canaan was a land that included much of present-day Lebanon and Israel.

Whatever language these people first spoke, Greek or something else, in time the Philistines apparently adopted a Canaanite tongue, for the Bible portrays them as having no trouble communicating with the Israelites. Phoenician and Hebrew were dialects of the Canaanite language.

But scholars have never found any unambiguous example of the writing of the Philistines, early or later.

A preliminary analysis of the inscription, Dr. Gitin said, showed that not only was the script Phoenician, but probably the language was as well. But it may have been a variation of Phoenician used by the Philistines, with differences on the order of those between British and American English.

A closer study of the inscription is being made by Dr. Gitin and Dr. Trude Dothan, an archaeologist at Hebrew University in Jerusalem who is the other leader of the Tel Miqne–Ekron project. They are being assisted by Dr. Joseph Naveh, a Hebrew University epigrapher, who was one of the scholars who originally suggested that the ruins might be those of Ekron.

Dr. Gitin said the inscription had already confirmed the close link between Ekron and the Neo-Assyrian Empire, which in the late eighth century B.C. and most of the seventh century B.C. was the superpower of what was then considered the known world.

Ekron was one of many vassal city-states in the empire and, as current excavations are revealing, must have been one of the largest industrial centers of any kind in the ancient Middle East in the seventh century B.C.

The name Achish in the text established the linkage for archaeologists. Achish was the name of a Philistine king mentioned in the Bible in the Books of I Samuel and I Kings during the time of King David and King Solomon of Israel. But he is not the Achish referred to in the inscription. Instead, the Achish in the text, archaeologists have determined, corresponds to the name Ikausu, who is mentioned in Assyrian annals of the seventh century B.C. as the King of Ekron.

Ikausu, scholars noted, was one of 12 kings of the Mediterranean coast called upon by the Assyrian king in the first quarter of the seventh century B.C. to provide building materials and their transport for the construction of a palace at Nineveh. Ashurbanipal, the successor, ordered the vassal kings of the Philistine cities, including Ikausu of Ekron, to support his military campaigns against Egypt.

The other name in the text—Padi, the father of Achish or Ikausu—is referred to in Assyrian documents at the time the empire's army conquered Ekron, which had been under the control of neighboring Judah. The Assyrians restored Ekron's status as a city-state, though now subservient to Nineveh, and reinstated Padi as its king.

The inscription thus documents a critical period in Ekron's history—its embrace by the Neo-Assyrian Empire and the expansion and apparent prosperity that followed.

The stone itself attests to the city's newfound wealth, for it celebrated the construction of a new temple on the west side of a stately palace, a building of Neo-Assyrian design and one of the largest structures of its kind to be excavated in Israel. Other digging in the last 13 years has shown that Ekron in the seventh century B.C. grew rapidly from not much more than 10 acres to a city of 85 acres, complete with an elite quarter in the center and an industrial zone containing more than 100 olive-oil processing plants.

These excavations are part of an ambitious study of the Neo-Assyrian Empire, especially its influence in the provinces and vassal city-states. The project is directed by the Albright Institute in Jerusalem, affiliated with the American Schools of Oriental Research, and Hebrew University and also involves a consortium of 22 North American and Israeli universities and research centers.

The discovery of the inscribed stone, Dr. Gitin said, "is going to allow us to write with a great deal of assurance the history of the Neo-Assyrian Empire and its revolutionary economic developments."

Through military might and political maneuvers, as well as innovative economic practices, the kings in Nineveh, the Assyrian capital on the Upper Tigris River in what is now northern Iraq, controlled territory as far south as Egypt and across present-day Syria, Iraq and parts of Turkey and Iran. The empire's Phoenician traders, operating out of the ports of Tyre and Sidon, extended a Syrian influence as far west as Carthage, Sicily and Iberia. Other economic links reached east into Afghanistan and perhaps India.

In their quest for raw materials and manufactured goods, as well as new sources of silver for use as currency, the Assyrian kings created a new supranational system of political and economic power, leading to 70 years of widespread growth of urban centers, transforming cottage industries into mass production and encouraging specialization in manufacturing. The heartland of the empire was extensively explored in the nineteenth century by European and American archaeologists who uncovered city ruins and royal documents. At the site of Ekron and in other research, scholars are concentrating on the view of the empire from the periphery, and Dr. Gitin is convinced that this is yielding a telling picture of the empire's dynamics.

In a report on his interpretations of Ekron's imperial role, published by the Archaeological Institute of America, Dr. Gitin wrote that the Philistine city "was apparently chosen as a focus of Assyrian economic activity because of its geographic and topographic advantages, with its proximity to sources of raw materials, land routes and Mediterranean harbors." In addition, the city had escaped serious damage in the Assyrian conquest and was a politically stable environment.

Archaeologists were particularly impressed by the extent of Ekron's olive-oil industry. In detailed excavations of only 3 percent of the city's area, they uncovered 105 olive-oil installations, containing stone presses, ceramic storage vessels and other artifacts.

When two such factories were reconstructed, researchers tested their output and determined that Ekron's estimated overall annual production of olive oil could have reached 1,000 tons, or 290,000 gallons. This is the equivalent of 20 percent of Israel's current level of export olive-oil production.

Before this time, archaeologists said, there is no evidence of olive-oil production in Ekron and very little elsewhere in the region, most of it for local consumption.

"This is a prime example of the innovative policy of industrial specialization and mass production which concentrated large-scale industrial activity in one center," Dr. Gitin said.

Other artifacts at Ekron pointed to a significant textile industry and to extensive foreign contacts, presumably through trade. Among the ruins are goblets and bottles from Assyria, ceramics from Greece and Carthage, and Israelite and Phoenician religious objects. And there are hoards of silver in small ingots and jewelry.

Another of the Neo-Assyrian innovations, it seems, was the widespread use of silver as a currency to supplement and, in some cases, replace conventional modes of payment by goods and services. In Spain, new silver mines were opened to meet the increased currency demands.

At the Ekron site, archaeologists came upon four large collections of silver, some hidden in cooking jugs buried beneath the floors and others found in a hole in a large stone—perhaps an early form of a wall safe.

Dr. Michael Notis, a metallurgist at Lehigh University, is analyzing the silver to determine its origin.

Other scholars praise the comprehensive excavations at Ekron and have generally endorsed Dr. Gitin's assessment of the innovative dynamics of the Neo-Assyrian Empire. But they cautioned against possibly exaggerating the role of Ekron in the empire, just because the research is new and in some cases surprising. Other cities, like Tyre and Sidon, were probably more important to the empire, they pointed out.

Ekron's time of prosperity was fleeting, as was the Neo-Assyrian Empire's. In the late seventh century B.C., first Egypt and then Babylon broke away from the empire, and Babylonian forces conquered Nineveh in 612. Ekron itself fell to the Babylonians of Nebuchadnezzar in 603, and the entire city and its grand palace with Achish's designatory stone became ruins.

Then the Philistines largely disappeared from history. The Neo-Assyrian Empire, which scholars consider the first of the classical empires, was followed by the Babylonians, Persians, Greeks and Romans.

—JOHN NOBLE WILFORD, July 1996

# Fortress Yields Clues to Israel's Ancient Foe

THE FAMILIAR BIBLICAL STORY of sibling rivalry, foreshadowing strife in the Middle East then and to this day, began when the pregnant Rebecca, wife of Isaac, was told by the Lord:

> *Two nations are in thy womb,*
> *And two manner of people shall be separated from thy bowels;*
> *And the one people shall be stronger than the other people;*
> *And the elder shall serve the younger.*
>
> —Genesis 25:23

Rebecca gave birth to twin sons, Esau and Jacob, whose lives fulfilled the prophecy. Esau, the firstborn, became a hunter who sold his birthright for a mess of pottage, a stew. Jacob tricked their dying father into granting a blessing that made Jacob master over Esau. The brothers, an angry Esau and a fearful Jacob, went their separate, antagonistic ways as patriarchs of rival nations.

Esau took wives from among the Canaanites and settled his family in the hill country of Seir, which became known as the land of Edom. This land, east of the Negev Desert of Israel, lies in present-day Jordan. Esau is called the father of the Edomites, about whom little is known except for the biblical accounts, which were colored by Israelite animosity.

New discoveries by archaeologists are beginning to cast light on Edom's shadowy history. Digging at several sites in Edomite country, notably Tell el-Kheleifeh, they have uncovered distinctive pottery from the first millennium B.C. and seals bearing inscriptions to Qos, the principal Edomite deity. Other inscriptions name one of the Edomite kings mentioned in Assyrian documents.

More recently, however, the discoveries are coming from ruins well within Israel itself and are revealing previously unknown aspects of Edomite religion and art in the seventh and sixth centuries B.C., apparently the height of the culture's political development. In a report in *Biblical Archaeology Review*, two Israeli archaeologists describe what they say is "one of the most spectacular finds of recent decades in all Israel."

At En Hatzeva, a site in the Negev 20 miles southwest of the Dead Sea, the archaeologists have discovered a hoard of religious artifacts, including shattered cultic figures that bear striking resemblances to Edomite material. Among the 75 objects are seven limestone incense altars, some stone sculptures with stylized human features, chalices and knee-high clay stands in the shape of human figures that were presumably used for making offerings to deities.

The artifacts were excavated about 50 feet from the ruins of a fortified city. The archaeologists suspect that this was a shrine from the seventh and early sixth centuries B.C. All the objects had been smashed, but every piece was there, indicating that they had been placed in a pit and destroyed by later iconoclasts. Most have now been restored.

"It is these anthropomorphic stands that most clearly suggest that the hoard may be Edomite," said the archaeologists, Dr. Rudolf Cohen and Dr. Yigal Yisrael of the Israeli Antiquities Authority. "The anthropomorphic figures probably represent priests or their followers, rather than gods."

Dr. Itzhaq Beit-Arieh, an archaeologist at Tel Aviv University, is less reserved in his judgment. "En Hatzeva clearly indicates an Edomite presence in the region connecting Edom and Judah." If he is right, the findings raise many questions. Is this evidence of an Edomite conquest at this time of part of the southern Israelite kingdom of Judah? Or is it just evidence that Edomite merchants set up an enclave there?

At the time, Judah was certainly vulnerable. In 586 B.C., the Babylonians had destroyed the temple of Solomon in Jerusalem. The northern kingdom of Israel was conquered by the Assyrians, and Judah was under increasing pressure from first the Assyrians and then the Babylonians. If the Edomites had indeed taken advantage of this weakness, as the research suggests, it would add an important but not surprising element to this chapter in Israel's history.

In other times, according to the Bible, the Edomites refused to allow the Israelites to pass through their territory on their way from Egyptian bondage to the Promised Land. Saul, the first king of Israel, waged war against Edom and his successor, David, conquered it. In Psalms 108:9, it is written, "Moab is my washpot; over Edom will I cast out my shoe; over Philistia will I triumph."

Whatever the truth of the story of Esau and Jacob, enmity between the two nations had been deep since they both emerged from obscure origins about 1200 B.C. "Because the origins of the Israelites and Edomites are so shrouded," Hershel Shanks, editor of *Biblical Archaeology Review,* said, "what we learn about one helps us understand the other."

Before the latest discoveries, the first evidence for possible Edomite incursions into a weakened Judah came from the ruins of Qitmit, 27 miles north of En Hatzeva. Dr. Beit-Arieh, who excavated Qitmit, identified it as an Edomite shrine on the basis of more than 800 artifacts, mainly figurines, human-shaped cultic stands and inscriptions referring to the Edomite god Qos.

At first, En Hatzeva held little promise as a possible Edomite site. It lay at a crossroads where pastoral nomads and caravans stopped for centuries. Earlier excavations there had yielded the remains of occupations in Roman, Byzantine and early Islamic times. Then ruins of a fortified city from the ninth to sixth centuries B.C. were also identified. It was outside one of these walls that Dr. Cohen, deputy director of the antiquities authority, and Dr. Yisrael, a specialist in Negev archaeology, came upon the buried remains of the shrine. They were particularly struck by similarities between the anthropomorphic stands there and at Qitmit. They were made on a potter's wheel, but limbs, noses, locks of hair and other details were modeled by hand. The stands are hollow cylinders, carry depictions of humans and animals and are topped with a bowl, presumably where the offering was placed, either burning incense or libations. Some are decorated with a pattern of triangles associated with artifacts found at undisputed Edomite sites. The architecture of En Hatzeva and Qitmit is also similar.

"Although the similarities between the sites are strong enough to suggest Edomite influence at En Hatzeva," Dr. Cohen and Dr. Yisrael wrote, "discrepancies among the two collections of artifacts make conclusive judgment difficult." The most they would say with certainty was that "the En Hatzeva shrine was dedicated to idol worship." But they cautioned that similar cult stands have turned up at Canaanite, Philistine and Phoenician sites.

In an article prepared for publication in *Biblical Archaeology Review*, Dr. Beit-Arieh concluded that the discoveries of considerable deposits of Edomite material in the Negev "indicates Edomite domination of the region at the end of the First Temple period."

This, in turn, could explain the ruins of many fortresses that have been found throughout the Negev. "Now it seems that this line of fortified sites and settlements was erected primarily against possible Edomite invasions," he said.

Although some may link this with Edomite threats and conquests in the region, Dr. Cohen and Dr. Yisrael doubt that these putative descendants of Esau had actually gained control of this part of the kingdom of Judah. "This influence, we believe, was largely economic," they wrote. "Whatever the ultimate answer, a shadow of uncertainty must linger over En Hatzeva for the time being."

The Edomites eventually faded from history. In the fourth century B.C., when the area was under Greek rule, Edom was referred to as Idumea, and by then it included much of the area of Judah that seems to have been under Edomite influence in the late seventh century or before. The mother of Herod, the Israelite king, was an Idumean, which was a kind of symbolic rejoining of the heirs of Esau and Jacob. The new discoveries may reveal even more linkages. Dr. Beit-Arieh said, "Like many peoples mentioned in the Bible but otherwise almost unknown, the Edomites are coming to life through archaeology. Ironically, however, some of the most dramatic finds are being excavated in Israel rather than in the Edomite homeland."

—John Noble Wilford, June 1996

# From Israeli Site, News of House of David

AN ISRAELI ARCHAEOLOGIST has discovered a fragment of a stone monument with inscriptions bearing the first known reference outside the Bible to King David and the ruling dynasty he founded, the House of David.

Scholars of biblical history said this was strong corroborating evidence for the existence and influence of the House of David in early Jewish history and in the traditions of both Judaism and Christianity. In their excitement, they used words like "phenomenal," "stunning" and "sensational" to emphasize the importance of the discovery in biblical archaeology.

The broken monument, or stele, was found in the ruins of a wall at Tel Dan, the site of an ancient city in northern Israel near the Syrian border and at one of the sources of the Jordan River. The discovery was made this summer by Dr. Avraham Biran, an archaeologist at Hebrew Union College and the Jewish Institute of Religion in Jerusalem and director of excavations at Tel Dan since 1966.

Dr. Biran said the stele was inscribed with 13 truncated lines of Aramaic text referring to the "House of David." From the style of the script and its references to a "king of Israel" and a king of the House of David, the archaeologist surmised that this probably was a victory stele erected in the first quarter of the ninth century B.C. by the king of Damascus after he "smote Ijon, and Dan, and Abel-beth-maachah," in the words from I Kings 15:20.

In that case, according to Dr. Biran's interpretation, the "king of Israel" of the inscription may be identified with Baasha and the king of the "House of David" with Asa, a descendant of David who ruled as king of Judah. A split among the Israelites after the death of Solomon in the tenth century B.C. had led to the northern kingdom of Israel and the southern kingdom of Judah, centered at Jerusalem. As related in I Kings, when war broke out between the two kingdoms, Asa secured an alliance with Ben-

Hadad, king of Aram at Damascus in Syria, who defeated the forces of Baasha.

In a telephone interview from Jerusalem, Dr. Biran said, "There has never before been found a reference to the House of David other than in the Bible."

Other scholars agreed, and noted that, for that matter, no reference to David himself had ever appeared before in nonbiblical texts. Indeed, as Dr. Jack M. Sasson, professor of religious studies at the University of North Carolina at Chapel Hill, said, "No personality in the Bible has been confirmed by other sources until Ahab, not David or Abraham or Adam and Eve."

King Ahab, husband of the notorious Jezebel, lived later in the ninth century B.C., dying in 897 B.C. David is supposed to have reigned from the year 1000 to 961. Dr. Sasson cautioned that the reference to the House of David did not necessarily prove the man existed. It could be, he said, that people who considered themselves his descendants had come to revere someone by that name who had been elevated to mythical standing. "Until you find a text actually written by David, people will wonder," he said.

Dr. Eric M. Meyers, a biblical archaeologist at Duke University, said, "It's a stunning discovery. Publication of the text should really enlighten us on the ninth century B.C., which has been a kind of dark age in biblical history."

Hershel Shanks, editor of *Biblical Archaeology Review,* said the findings provided contemporaneous evidence supporting accounts of the Jewish monarchies in I Kings and II Chronicles. He said, "The stele brings to life the biblical text in a very dramatic way. It also gives us more confidence in the historical reality of the biblical text—in a broad way, not necessarily in regard to each detail."

Although Dr. Biran gave a terse description of his findings in correspondence with colleagues this week, he said photographs and transcriptions of the writing on the stele would not be released until a full report was ready for publication in the *Israel Exploration Journal.* Dr. Joseph Naveh, an epigrapher of ancient Semitic languages at Hebrew University in Jerusalem, is in charge of deciphering and analyzing the text.

Dr. Biran said that the fragment perhaps represented only one third of the stele, which he estimated to have been at least three feet high. Despite the many gaps in the Aramaic text, he said, "the writing is very clear, a joy to behold."

There could be no mistaking the words for "House of David"—"bet David," the same in Aramaic and Hebrew. As Dr. Biran noted, periods separate each word in the text, except for the two words for House of David. It was as if to emphasize the special status of the expression, he said, and thus to indicate the prominence of the dynasty in the political affairs of the day.

The earliest remains uncovered at Tel Dan indicate that the site has been occupied more or less continuously since the fifth millennium B.C. The Israelite tribe of Dan apparently settled there in the twelfth century B.C. The stele fragment was found near a gate along the southeastern border of a stone-paved piazza. The excavations are being conducted with the support of the Israel Antiquities Authority.

Dr. Biran surmised that in a reversal of fortune the kingdom of Israel defeated the Arameans in another war some 30 years after the erection of the stele at Dan. The victorious King Ahab apparently had Ben-Hadad's old stele shattered. In time the fragments were used in the construction of the pavement and surrounding walls. Archaeologists are continuing the search, hoping to find the rest of the stele.

If Dr. Biran's preliminary analysis of the text is confirmed, Dr. Meyers said this would support biblical accounts of some of the political events after the tribes of Judah and Benjamin split off from the 10 other tribes of Israel to form the southern kingdom of Judah, ruled by successors of David. Modern Jews trace their lineage from this kingdom, and St. Luke states that Jesus was "of the house and lineage of David."

The other tribes formed the kingdom of Israel in the north. Scholars do not know what happened to these tribes after the end of the eighth century B.C., when they were taken into exile by the Assyrians and became the "lost tribes of Israel" in legend and speculative history.

The stele text, Dr. Meyers said, appears to deal with the time when "the king of Judah, Asa, raided the temple treasury to pay off the king of Syria to beat up on the king of Israel."

In I Kings, it is written that Asa "took all the silver and the gold that were left in the treasures of the house of the Lord" and delivered them to Ben-Hadad in Damascus as an inducement for him to break his treaty with

Baasha, king of Israel, and help Asa defeat this rival king. The stele also makes important references to the Aramean god of storms and warfare, Hadad, and to chariots and horsemen presumably captured by Ben-Hadad from Baasha.

With this help, Asa enjoyed a long reign and died of a festering foot disease. In II Chronicles 16:12–13, it is written that "in his disease he sought not to the Lord, but to the physicians." Thus, "Asa slept with his fathers, and died in the one and fortieth year of his reign." He was succeeded by a son, Jehoshaphat.

—JOHN NOBLE WILFORD, August 1993

# Biblical Puzzle Solved: Jerusalem Tunnel Is a Product of Nature

UNDER THE OLDEST PART of Jerusalem, the area known as the City of David, a maze of tunnels and shafts runs through the rock and deep into biblical history. In ancient times, the people inside the city walls depended on this system to deliver water from the ever-flowing Gihon Spring outside, ensuring a dependable water supply in war and peace.

But nearly everything else about the old underground waterworks, especially its recorded role in two pivotal events in the history of ancient Israel, has left scholars shaking their heads in puzzlement.

Archaeologists and biblical scholars have long wondered if it was these dark, subterranean passages that enabled King David to capture Jerusalem 3,000 years ago. Biblical accounts suggest that David's general, Joab, surprised the Jebusites, or Canaanites, by sneaking in through a hidden passage. But did any of these tunnels exist at this early time? Were the Canaanites or anyone else then capable of such excavations?

Engineers have long noted that whoever built these passages seemed to go about the task in the most curious way, with no logic in the choice of some routes, slopes and dimensions of the tunnels and many ostensible mistakes in design.

Take Hezekiah's Tunnel. According to the Bible, King Hezekiah, expecting an attack and possibly a long siege by the Assyrians in the eighth century B.C., had a tunnel built to bring water from the spring to an open reservoir within the walled city, which extends south of the Temple Mount. The siege occurred in 701 B.C., but failed, presumably in no small part because of the tunnel and its secure water supply. But why did the tunnel follow such a serpentine course, extending 1,748 feet, when a straight line of 1,050 feet would have been sufficient and easier to build?

Now many of these questions can apparently be answered. Previous explanations had been based on the assumption that the tunnels were entirely man-made. Scholars should have consulted a geologist sooner.

A comprehensive geological study of underground Jerusalem has recently shown that the channels and shafts were formed by natural forces tens of thousands of years ago. That means there may have been an underground passage through which Joab infiltrated the Canaanite city. And Hezekiah's Tunnel is winding and irregular because the builders simply modified a natural fissure.

Dr. Dan Gill, a senior geologist with the Geological Survey of Israel, first reported the discovery in 1991 in the journal *Science.* Underlying the City of David, he found, is a well-developed karst system, a geological term for a region of irregular sinks, caverns and channels caused by groundwater seeping through underground rock, mainly limestone and dolomite. Similar processes account for the many caves under the limestone of Kentucky.

In *Biblical Archaeology Review,* Dr. Gill has described the findings in more detail and discussed their implications for archaeological research and biblical history. The geology, he said, provides "a simple, consistent and unified solution" to "most of the puzzles that have heretofore stumped researchers."

The extent and peculiarities of the underground water system were discovered and explored in the nineteenth century. The passages were all connected to Gihon Spring, the Old City's sole source of fresh water and the reason that a city came to be built there. Modern Jerusalem's water supply is piped in from Lake Tiberias.

From Gihon Spring, which is in a cave, there runs a short, irregular tunnel leading to a vertical shaft that goes straight up 37 feet. This is called Warren's Shaft, after the British engineer Charles Warren, who explored it in 1867.

Someone standing on a rock platform at the top of the shaft could drop a bucket on a rope and draw up the cool water. A gently sloping tunnel, and then a steeper one, connect the platform with an entryway at the surface. Though the spring is a little outside the wall, the entryway to Warren's Shaft is safely inside.

Another important component, Hezekiah's Tunnel, was rediscovered in 1837 by Edward Robinson, an American Orientalist. The tunnel, drawing on the same spring, runs from the base of Warren's Shaft until it debouches in an open reservoir known as the Pool of Siloam. An inscription on the tunnel wall, written in ancient Hebrew script, tells how two teams digging from opposite ends managed to meet in the middle. That was an achievement that scholars found virtually inexplicable because of the winding route the tunnel followed, but the new findings show that the workers were actually following and widening the route of existing passages.

Systematic explorations were not renewed until 1978, when the late Dr. Yigal Shiloh, an Israeli archaeologist, began research on the City of David. Dr. Gill, as the project's chief geologist, reexamined the waterworks and in 1980 began to recognize a clear example of function following form.

Beneath the City of David, he found, lie two layers of rock, highly porous limestone on top of more impervious dolomite. Warren's Shaft is a natural sinkhole that developed along a joint in the limestone. Its bottom narrows into a funnel-like shape, typical of a karstic sinkhole, and carbon dating of the calcium crust on its walls indicates an age of more than 40,000 years.

"This provides unequivocal evidence that the shaft could not have been dug by man," Dr. Gill wrote.

But the inhabitants of ancient Jerusalem probably took a hand to make the sinkhole into a well, he said, by widening the fissure between it and the spring and by sealing the bottom of the shaft to prevent leakage. Likewise, the people modified natural passages from the top of the shaft to the surface, widening them and cutting out steps.

David's capture of Jerusalem around 1000 B.C. is described in the books of II Samuel and I Chronicles. Responding to Canaanite taunting that he could not conquer their city, David instructed Joab to have his men reach, or perhaps climb, the *tsinnor.* The use of this word, variously translated as water shaft, battlement, scaling hook or even penis, has led to no end of speculation.

After a careful analysis of the word's use elsewhere in the Bible and in classical Hebrew, Dr. Terence Kleven, professor of Jewish history at Hebrew University in Jerusalem, has concluded that *tsinnor* should be translated as a conduit for water or a water shaft.

But most archaeologists, including Dr. Shiloh, had not thought the Canaanites were capable of constructing the kind of underground system found at the City of David. Such man-made waterworks are unknown in the pre-Israelite cities of the area.

Linguistics and archaeology aside, Dr. Gill has now concluded, the new evidence established that Jerusalem in the time of the Canaanites could have been entered through at least two underground passages, the one at the cave of Gihon and another, narrow tunnel higher on the hillside, which is now blocked. Both tunnels connect to Warren's Shaft, the geologists said, and even before they were modified as a water system, they were probably wide enough for a person to pass through.

Although dolomite is more solid than limestone, cracks do occur under seismic stresses or at boundaries where different layers meet, and over time the erosion of water seeping along the fissures can leave substantial horizontal passages, Dr. Gill explained. This could explain most of the anomalies in Hezekiah's Tunnel, especially its serpentine route and the varying height of its ceiling from 6 feet to as much as 16. It could also explain how the workers, using oil-burning lamps, could get enough air to survive the construction.

One engineering modification was necessary to the success of the tunnel. The floor had to be cut to provide a slight descending slope, a mere 12.5 inches in the course of 1,748 feet, so that water could flow from the spring to the reservoir.

"This precision," Dr. Gill wrote, "suggests that anomalies like the varying height of the ceiling and the circuitous route of the tunnel were not the result of incompetence or carelessness."

In a telephone interview from Jerusalem, Dr. Gill said the "evidence is very, very clear" that the waterworks of ancient Jerusalem were created by the skillful adaptation of these natural fissures, providing a plentiful form of water, and this was what may have saved Jerusalem from the Assyrians.

—JOHN NOBLE WILFORD, August 1994

# Seeking a Replica of the Second Temple

"THIS IS WHERE WE FOUND a wonderful thing," Dr. Yitzhak Magen said, almost bubbly as he picked up a trowel and dug into a small mound of ash. Within seconds, he had scooped up a handful of charred bones of yearling sheep and goats killed more than 2,100 years ago.

"We think that this is where they did their sacrifices," said Dr. Magen, the Israeli government's chief archaeologist for the West Bank. "We found thousands of bones here."

He was standing by a pile of stones that he believes formed an altar of an ancient Samaritan temple built on the craggy top of Mount Gerizim, venerated by the tiny Samaritan community that endures here just south of the Palestinian town of Nablus. Finding remnants from that temple is satisfying enough, Dr. Magen says. But the discovery may have added poignancy.

He is convinced, based on the writings of the first-century historian Flavius Josephus, that the structure here was a replica of the Second Temple of Jerusalem, the core of Jewish life from the time its construction was started in 520 B.C. until, after many remodelings, it was destroyed in A.D. 70.

Before anyone conjures up Indiana Jones visions of a lost ark lying beneath Mount Gerizim's rocky soil, Dr. Magen points out that the temple and the city surrounding it were burned to the ground in 113 B.C. by the army of John Hyrcanus, leader of the Hasmonean rulers in Judea. But having found remains like the altar, he is certain that there is more, though there is not likely to be much more, he says. His goal is to find whatever else may be left under the ruins of the Church of Mary Theotokos, which was built on the site starting in A.D. 484.

"We believe that it was here, exactly, that the temple stood," he said, walking across the exposed pavement stones of the Byzantine church.

Josephus wrote that the Samaritan temple was built in the late part of the fourth century B.C., although Dr. Magen suspects that work may actually have started dozens of years later. According to Josephus, there was a love story behind the construction.

Manasseh, a Jewish high priest in Jerusalem, went against his people's traditions by marrying Nikaso, who was a Samaritan and, as such, a member of a sect that had many customs similar to those of Judaism but that was a bitter rival of the Jews for centuries. Essentially, temple elders gave Manasseh two choices: Give up his wife or leave the temple. He chose her. But Nikaso's father, Sanballat, who was a Samaritan leader, promised to ease the pain of that decision. He built a temple on Mount Gerizim modeled on the one in Jerusalem, Josephus said, with Manasseh installed as chief priest.

Dr. Magen and his research teams have been excavating the 2,850-foot peak since 1983, but it is only in the last few years that outer precincts of the temple have emerged.

One find was the altar, made of uncarved stone as specified in the Pentateuch, the first five books of the Old Testament. The Samaritans rigorously follow the Pentateuch while rejecting the later prophets and the oral traditions that form the basis of Jewish rabbinical law. The altar is near a stone entrance that Dr. Magen says formed the temple's northern gate, more than 65 feet wide.

"If it is that big," he said, "you can imagine the size of the temple enclosure."

His crew, made up of several dozen Palestinians from nearby Nablus, has unearthed what seem to be other gates facing east and west. On the eastern end of the dig, the workers found a series of walls, one atop another—Byzantine era above Hellenistic era above a long stretch of stone running perhaps 450 feet that Dr. Magen believes may have been the temple wall.

To the west is a staircase of seven stone steps that seem to lead to the temple compound. To the south, west and north, archaeologists have uncovered the ruins of 20 stone houses, a small part of an ancient city of some 10,000 inhabitants spread across more than 100 acres. It is believed, Dr. Magen said, that priests and other temple workers lived there, "close to the temple itself just as their Jewish counterparts did in the Second Temple area."

Three-foot piles of ash found by the diggers attest to the destruction inflicted by John Hyrcanus' forces. But enough of the houses remain to make it clear that they had two floors, and each had its own bathroom, with stone tubs still intact in some. Only a few hundred yards away, within easy sight of these ancient living quarters, are the squat concrete houses of modern Samaritans, a faded community of fewer than 600 people who live both on the lower reaches of Mount Gerizim and in Holon, a working-class suburb of Tel Aviv.

One bit of good fortune that has surfaced in recent months is a stone fragment from the second or third century B.C. that was inscribed with the Ten Commandments. It was written in the Samaritan script, which is similar to an ancient form of Hebrew known as Paleo-Hebrew.

Other fragmentary inscriptions found on the site contain the word *cohen,* or priest; its plural form, *cohanim;* the name Pinhas, which might refer to a Samaritan priest; and equivalents to the letters Y H V H—the abbreviation for Jehovah, God's name, which Jews are forbidden to pronounce.

Taken together, the finds make Dr. Magen certain that he is on the right track to the heart of the Samaritan temple itself, just below the Mary Theotokos church. Thus far, excavations have been confined largely to the edges of the church, a broad open-air compound measuring 235 feet by 200 feet that itself was mostly covered until the Israeli archaeologist began his work.

"Centuries after the temple was burned by Hyrcanus and they were banished, the Samaritans came back," he explained. "This was after the Roman era and before the Byzantine period starting in the fourth century. When they came back, they built synagogues and other buildings, and we want to isolate all that before going to the temple area itself.

"It's like a puzzle, and you have to put the pieces together," he said. "It takes a lot of luck."

Dr. Magen, who works for the Civil Administration, which despite its name is the Israeli military government for the West Bank, hopes to begin probing beneath the stone church floor within a few months.

Inevitably, the tantalizing question is whether he will find clues to what the Second Jewish Temple in Jerusalem looked like.

The First Temple, begun in 960 B.C. under King Solomon, was destroyed by the Babylonians in 586 B.C. It was only after the return of the Jews from Babylonian exile in 538 B.C. that work on the Second Temple began.

Ancient writings describe it as more modest than the original, but there was no clear description of its construction or layout, as there was for Solomon's temple in the First Book of Kings. Nonetheless, references to it in the books of Ezra and Nehemiah and in postbiblical commentaries suggest that there were two courtyards with chambers, gates and a public square. In the late first century B.C., after the Samaritan copy was destroyed, King Herod I of Judea built a far more splendid structure on the original site, at the same time expanding the boundaries of the Temple Mount on which the edifice sat. What remains of all that is only a section of the western supporting wall of the Temple Mount, commonly known as the Wailing Wall, or Western Wall, for many Judaism's holiest site.

Despite the sketchiness of details, there have been attempts by artists over the years to show how the early Second Temple might have looked.

Is there a chance that Dr. Magen will find solid confirming evidence beneath the church on Mount Gerizim? Probably not, he says, given the destruction in the second century B.C.

But he says he will try anyway. He also recognizes that he may be in a race against time.

Israel and the Palestine Liberation Organization are negotiating an expansion of the fledgling Palestinian self-rule beyond its present confines of the Gaza Strip and the small West Bank district of Jericho. Nothing is settled yet. But it is possible that Nablus, with other parts of the West Bank, will fall under Palestinian authority. And if arrangements for the last year in Gaza and Jericho are a guide, it means that responsibility for archaeological works will also come under Palestinian control.

It is always possible that the Palestinians will continue the excavations here, but that seems highly unlikely considering both the expense and the displeasure that the Samaritan community has already expressed about the digging on their sacred mountain. Besides, Palestinian officials have shown scant interest in any finds that show Jewish roots in territories they consider their own.

So could politics indeed outpace archaeology? Dr. Magen says he will let others worry about it.

"We have an expression in Yiddish—*Kullu b'id Allah,*" he said. It was a small joke. He was speaking in Arabic: Everything is in God's hands.

—CLYDE HABERMAN, May 1995

# 5

# THE NEW WORLD

The archaeology of the New World focuses on the great pre-Columbian civilizations of Mesoamerica and the Andes as well as such questions as the date of the first human occupants and the earliest domesticated crops.

The pre-Columbian civilizations appear to have developed independently from those of the Old World and so exhibit an alternative pathway in cultural development. As in the Old World, sophisticated societies developed, with art and architecture, a writing system and, in the case of the Maya, an elaborate calendar based on astronomical observation.

Blood, sacrifice and the merciless treatment of captives seemed to have played a central role in Mayan culture, or at least in the sample apparent in the archaeological record. The reasons for the culture's collapse remain a matter of scholarly dispute.

# Human Presence in Americas Is Pushed Back a Millennium

AFTER LONG, OFTEN BITTER DEBATE, archaeologists have finally come to a consensus that humans reached southern Chile 12,500 years ago. The date is more than 1,000 years before the previous benchmark for human habitation in the Americas, 11,200-year-old stone spear points first discovered in the 1930s near Clovis, New Mexico.

The Chilean site, known as Monte Verde, is on the sandy banks of a creek in wooded hills near the Pacific Ocean. Even former skeptics have joined in agreeing that its antiquity is now firmly established and that the bone and stone tools and other materials found there definitely mark the presence of a hunting and gathering people.

The new consensus regarding Monte Verde represents the first major shift in more than 60 years in the confirmed chronology of human prehistory in what would much later be called, from the European perspective, the New World.

For American archaeologists it is a liberating experience not unlike aviation's breaking of the sound barrier; they have broken the Clovis barrier. Even moving back the date by as little as 1,300 years, archaeologists said, would have profound implications for theories about when people first reached America, presumably from northeastern Asia by way of the Bering Strait, and how they migrated south more than 10,000 miles to occupy the length and breadth of two continents.

It could mean that early people, ancestors of the Indians, first arrived in their new world at least 20,000 years before Columbus.

Evidence for the pre-Clovis settlement at Monte Verde was amassed and carefully analyzed over the last two decades by a team of American and Chilean archaeologists, led by Dr. Tom D. Dillehay of the University of Kentucky in Lexington.

Remaining doubts were erased by Dr. Dillehay's comprehensive research report, which has been circulated among experts and published by the Smithsonian Institution.

A group of archaeologists, including some of Monte Verde's staunchest critics, inspected the artifacts and visited the site, coming away thoroughly convinced. In his report of the site visit, Dr. Alex W. Barker, chief curator of the Dallas Museum of Natural History, said: "While there were very strongly voiced disagreements about different points, it rapidly became clear that everyone was in fundamental agreement about the most important question of all. Monte Verde is real. It's old. And it's a whole new ballgame." The archaeologists made the site inspection under the auspices of the Dallas museum, where their conclusions were reported, and with additional support by the National Geographic Society.

The archaeologists, all specialists in the early settlement of America, included Dr. C. Vance Haynes of the University of Arizona, Dr. James M. Adovasio of Mercyhurst College in Erie, Pennsylvania, Dr. David J. Meltzer of Southern Methodist University in Dallas, Dr. Dena Dincauze of the University of Massachusetts at Amherst, Dr. Donald K. Grayson of the University of Washington in Seattle and Dr. Dennis Stanford of the Smithsonian Institution in Washington.

Dr. Dincauze, who had expressed serious doubts about the site's antiquity, said that Dr. Dillehay's report made "a convincing case" that the remains of huts, fireplaces and tools showed human occupation by a pre-Clovis culture.

"I'm convinced it's one hundred percent solid," Dr. Brian M. Fagan, an anthropologist at the University of California at Santa Barbara, said of the new assessment of Monte Verde. "It's an extraordinary piece of research."

Finally vindicated, Dr. Dillehay said, "Most archaeologists had always thought there was a pre-Clovis culture out there somewhere, and I knew that if they would only come to the site and look at the setting and see the artifacts, they would agree that Monte Verde was pre-Clovis."

Monte Verde, on the banks of Chinchihaupi Creek, is in the hills near the town of Puerto Montt, 500 miles south of Santiago. Dr. Dillehay and Dr. Mario Pino of the Southern University of Chile in Valdivia began excavations there in 1976. They found the remains of the ancient camp, even wood and other perishables that archaeologists rarely find, remarkably well preserved by the water-saturated peat bog that covered the site, isolating the material from oxygen and thus decay.

As Dr. Dillehay reconstructed the prehistoric scene in his mind, a group of 20 to 30 people occupied Monte Verde for a year or so. They lived in shelters covered in animal hides. They gathered berries in the spring, chestnuts in the fall and also ate potatoes, mushrooms and marsh grasses. They hunted small game and also ancestors of the llama and sometimes went down to the Pacific, 30 miles away, for shellfish. They were hunters and gatherers living far from the presumed home of their remote ancestors, in northeastern Asia.

The evidence to support this picture is extensive. Excavations turned up wooden planks from some of the 12 huts that once stood in the camp, and logs with attached pieces of hide that probably insulated these shelters. Pieces of wooden poles and stakes were still tied with cords made of local grasses, a telling sign that ingenious humans had been there. "That's something nature doesn't do," Dr. Barker said. "Tie overhand knots."

Stone projectile points found there were carefully chipped on both sides, archaeologists said. The people of Monte Verde also made digging sticks, grinding slabs and tools of bone and tusk. Some seeds and nuts were shifted out of the soil. A chunk of meat had managed to survive in the bog, remains of the hunters' last kill; DNA analysis indicates the meat was from a mastodon. The site also yielded several human coprolites, ancient fecal material.

Nothing at Monte Verde was more evocative of its former inhabitants than a single footprint beside a hearth. A child had stood there by the fire 12,500 years ago and left a lasting impression in the soft clay.

Radiocarbon dating of bone and charcoal from the fireplaces established the time of the encampment. The date of 12,500 years ago, said Dr. Meltzer, author of *Search for the First Americans,* published in 1993 by the Smithsonian Institution, "could fundamentally change the way we understand the peopling of the Americas."

The research, in particular, shows people living as far south as Chile before it is clear that there existed an ice-free corridor through the vast North American glaciers by which people might have migrated south. In the depths of the most recent ice age, two vast ice sheets converged about 20,000 years ago over what is now Canada and the northern United States and apparently closed off human traffic there until sometime after 13,000 years ago. Either people migrated through a corridor between the ice sheets and spread remarkably fast to the southern end of America or they came by a different route, perhaps along the western coast, by foot and sometimes on small vessels. Otherwise they must have entered the Americas before 20,000 years ago.

Dr. Carol Mandryk, a Harvard University archaeologist who has studied the American paleoenvironment, said the concept of an ice-free corridor as the migration route emerged in the 1930s, but her research shows that even after the ice sheets began to open a path, there was not enough vegetation there to support the large animals migrating people would have had to depend on for food.

"It's very clear people couldn't have used this corridor until after thirteen thousand years ago," Dr. Mandryk said. "They came down the coast. I don't understand why people see the coast as an odd way. The early people didn't have to be interior big-game hunters, they could have been maritime-adapted people."

No archaeologists seriously consider the possibility that the first Americans came by sea and landed first in South America, a hypothesis made popular in the 1960s by the Norwegian explorer Thor Heyerdahl. There is no evidence of people's occupying Polynesia that long ago. All linguistic, genetic and geological evidence points to the Bering Strait as the point of entry, especially in the Ice Age, when lower sea levels created a wide land bridge there between Siberia and Alaska.

Although several other potential pre-Clovis sites have been reported, none has yet satisfied all archaeologists in the way Monte Verde has done. But archaeologists expect that the verification of Monte Verde will hasten the search for even older places of early human occupation in the Americas.

—John Noble Wilford, February 1997

# Cave Filled with Glowing Skulls: A Pre-Columbian Palace of the Dead

THE NARROW ROAD into the Honduran rain forest was a mire of black mud, a grinding test for the hardiest four-wheel-drive vehicles. After fording a wild river, everyone had to get out and hike up a steep slope. The rocky trail was slippery, crawling with fire ants and bounded by green walls of vines hanging from tall trees.

This was the way to Cueva de R o Talgua, the Cave of the River Talgua, a haunting place the explorers had taken to calling the Cave of the Glowing Skulls. It is the site of a newfound archaeological mystery.

A tongue of water rushed out of the cave's dark mouth, tumbling noisily over rocks and down the hill to the Talgua. Following the stream, sometimes wading up to their thighs, archaeologists plunged several hundred yards into the interior for their first scientific examination of a discovery made in April 1994. They ventured through side passages and up into chambers well above stream level. They finally passed through a small opening near the ceiling of one chamber and by the light of their headlamps, caught their first sight of an astonishing scene.

It was a pre-Columbian palace of the dead.

Stalactites of calcium carbonate, calcite, dripped from the craggy ceiling of a cavern more than 100 feet long, 12 feet wide and as much as 25 feet high. The timeless seepage of water through limestone had left deposits of calcite everywhere, seemingly frozen in midflow. The walls gleamed like marbleized ice. Hanging at angles to the walls were ceiling-to-floor curtains of calcite, rigid, delicately thin and almost translucent. They looked like acoustic wings in an Art Deco concert hall.

In the recesses of the curtains, in every crevice and on every ledge, were piles of human skulls and bones, sparkling with coatings of tiny calcium crystals.

Kneeling in one corner, Dr. James E. Brady, an archaeologist at George Washington University and leader of the investigation, spoke with growing excitement. "Look here, two, four, five, six skulls," he said. "Here's some red pigment, something often associated with burials as far back as the Neanderthals. Who knows how many more bones are beneath these, cemented in the calcite? This place is just filled with an incredible quantity of human bone."

After several more daylong visits to the cavern, Dr. Brady would estimate that the visible remains represented from 100 to 200 individuals. Who were these people, to what ancient culture did they belong? When did they live and die? These are questions that puzzle and intrigue archaeologists as they pore over the few clues and plan radiocarbon tests of some ash samples and bone fragments.

Dr. Brady, who specializes in Maya cave archaeology, said the evidence so far ruled out any close relationship between these people and the Maya, whose civilization dominated upper Central America and southern Mexico in the first millennium A.D. But the 20 undecorated ceramic bowls and two thin marble vases found with the bones could not be matched with the styles of known non-Mayan cultures in what is now Honduras. Based largely on the ceramics, he said, the burials could have occurred as recently as A.D. 500 or as early as 300 B.C.

"It's frustrating," Dr. Brady sighed. "We have all this beautiful material and no way of immediately relating it to a certain time."

A cow pasture less than a mile from the cave entrance may hold answers. George Hasemann, director of archaeology at the Honduran Institute of Anthropology and History in Tegucigalpa, who accompanied Dr. Brady, identified more than 100 large rectangular mounds in the pasture, presumably the remains of an ancient settlement. Some pottery shards recovered there were similar to those found in the cave, he said, indicating that the site was occupied at the same time the cave was being used as an ossuary.

As the next step in trying to solve the mystery of the Talgua cave, Mr. Hasemann proposed a thorough reconnaissance of the mounds, beginning with some test trenches. Such exploration, combined with the cave findings, could open a window on a previously unknown culture that lived in the shadow of the mighty Maya civilization.

"This is a tremendously important find," said Dr. Rosemary Joyce, director of the Phoebe Hearst Museum of Anthropology at the University of California at Berkeley, a specialist in Honduran archaeology. "We're dealing with a geographical area about which almost nothing is known archaeologically."

What is known is that the area of central and northeast Honduras was a heavily populated region on the periphery of the Maya cultures. The people of this region undoubtedly had some contact with the Maya, particularly with the great city of Copan in northwest Honduras. Indeed, Dr. Joyce speculated, these people, perhaps including those interred in the cave, could have been middlemen in cultural traffic between lower Central America and the Maya world. Through them may have passed the goldworking technologies of Panama and Costa Rica to the Maya and the jade crafts of the Maya to the rest of Central America.

Dr. Joyce has not visited the site, but has examined the pottery. This led her to give an even earlier estimate for the cave burials—as early as 800 B.C. or as late as 100 B.C.

The Talgua cave is about four miles northeast of Catacamas, a rough-and-ready farm town where some men openly carry pistols and machete mayhem is not uncommon. It is near the fringes of the Mosquitia Rain Forest and 100 miles northeast of Tegucigalpa, the capital.

People in town knew about the cave and had ventured inside, leaving graffiti on the walls of the main passage. Surveyors had mapped the two-mile extent of the cave's stream level, but had overlooked the opening to the skull-filled inner sanctum.

"I'm not surprised that they didn't find the opening," said John H. Fogarty, a speleologist from Austin, Texas, who mapped the skull cavern for Dr. Brady. "It's a nasty hard climb to a little hole."

In April 1994 four young men from Catacamas, two Hondurans and two Americans, made their way to the back chambers, saw the opening and were overcome with curiosity. One, Jorge Yanez, went to town and got a tall ladder. He was the first to glimpse the ossuary. Fortunately, the spelunkers took pictures, left the chamber largely undisturbed and reported the discovery to Mr. Hasemann.

Archaeologists have to take their discoveries and allies as they come. It happened that at the time Steven E. Morgan, a Californian who describes himself as an explorer and treasure hunter, was in Honduras filming a 2,000-year-old burial site. Mr. Hasemann said he knew and trusted Mr. Morgan and asked him to join in a preliminary investigation of the cave in May. Impressed with what they saw, they reached an agreement in which Mr. Morgan could make a movie of the discovery if he allowed archaeologists to accompany the expedition and have first access to the skull chamber.

With the help of Laurence August, a New Orleans public relations man with Honduran clients, Mr. Morgan raised money and got gifts of equipment for the expedition to what he called the Cave of the Glowing Skulls. Archaeologists winced at the appellation, and acknowledged that the arrangement had its stresses and strains, but felt they had little choice.

"We wouldn't be here doing archaeology if it hadn't been for the filming," Dr. Brady said. "There's little money for archaeology, especially in Honduras and for small projects that are as speculative as this was. We just had to get archaeologists into the cave fast, once word of the discovery got out."

Mr. Fogarty said he had been in larger and more spectacular caves, but this one was fairly easy to work in. The environment was benign. Bats inhabited only the lower stream level, and the commotion had put them to flight. Nothing creepy and crawly seemed to be about. Most of the passages and chambers were expansive enough to allow the explorers to stand upright and avoid claustrophobia. Except for some dripping and seepage, the upper levels were reasonably dry, perhaps because of the summer drought.

Working in the cave, Dr. Brady saw enough to make some preliminary observations. Most of the long bones were in stacks, not laid out as normal skeletons. These were secondary burials, he concluded. The bodies had originally been buried elsewhere, until all flesh decayed. Then the bones were gathered in bags and taken into the cave for their permanent resting place.

One can imagine the ceremony. People would have walked by flickering torchlight through the dark passages, up two levels and into the remote chamber, bearing bundles of their ancestors' bones. They placed the bones in the wall recesses, perhaps laying them in pools of standing water. They may have left offerings in the bowls. They daubed the bones with a paint of red minerals, the color of blood, a last blush of life.

The archaeologists could not tell if these were elite burials. The artifacts were too few; no jade, no goods that might suggest high status and no evidence that the good stuff had been looted. Some skulls were deformed into the shape of truncated cones, the result of bindings in infancy and sometimes a mark of the elite.

"The Maya did this all the time," Dr. Brady said. "The practice of skull deformation may have come from the Maya, or it may be something that was general all over lower Central America. We just don't know yet."

It may be that these were burials of a particular kin group, which was sometimes the case among the Maya. An analysis of the bones may provide a clue. Some bone fragments will be examined for residues of collagen, the protein that holds together minerals of the bone. If there is any collagen, the archaeologists want to see if they can use it to date the bones through radiocarbon tests and do DNA studies, which might establish kinship.

Dr. Brady has investigated Maya caves at Dos Pilas in Guatemala and elsewhere. Among the Maya and most pre-Columbian societies, he said, the earth was seen as an animate and sacred entity. Caves represented the essence of their religious beliefs because they penetrated the sacred earth. In their creation stories, people emerged from caves, especially the god-kings. It was perhaps natural that in death they should want to be returned to their origins, in caves.

At Dos Pilas, caves were so sacred that the Maya built their temples directly over them to imbue the temples with that sense of sacredness. As recently as the early twentieth century, Dr. Brady said, Maya descendants in some rural areas were still placing their dead in caves.

"But here," Dr. Brady said, interrupting that train of thought, "we're outside the Maya area. I don't want to impose a Maya imprint on what we find, simply because I don't know who these people were and what this all means."

Still, he could not contain his surprise and joy over the discovery.

"All the time people come to me with stories about caves," he said. "They tend to be overly impressed with what they see. But I'm generally disappointed by what I find. In this case, I was overwhelmed with the quantity of skeletal material and the marble vases. If we only knew who these people were."

—JOHN NOBLE WILFORD, October 1994

# Age of Burials in Honduras Stuns Scholars

A laboratory analysis of material from a Honduran cave filled with human skulls and bones has produced two potentially significant surprises.

People were burying their dead in the cave as long ago as 3,000 years, about the time King David was capturing Jerusalem and before the founding of Rome. This was a considerably earlier time than previously estimated and means that the cave holds the earliest scientifically dated evidence for the emergence of complex society in Honduras. It may be that only the Olmec society of southern Mexico is older in Central America.

The humble beginnings of the great Maya city Copan in northwestern Honduras probably came a century or so later. The classic period of the Maya civilization throughout upper Central America and southern Mexico was in the first millennium A.D.

Even more puzzling to archaeologists than the early date was an analysis of bones indicating that the people buried in the cave apparently had not eaten corn. This pre-Columbian dietary staple was domesticated in the tropical lowlands of Mexico about 7,000 years ago, and scholars have generally linked cultivated corn with the rise of complex societies in much of the Americas.

The results of these studies were described by Dr. James E. Brady, an archaeologist at George Washington University in Washington, who has led the scientific investigation of the Cave of the River Talgua near the edge of a rain forest. He is a specialist in Maya cave archaeology.

At Dr. Brady's direction, Beta Analytic Inc. of Miami ran radiocarbon tests on two charcoal samples found among the skulls and bones. The first sample, which was also associated with a ceramic vessel, was dated at about 800 B.C., plus or minus 20 years. The second ash sample was even older, dated at 980 B.C., with a range of 905 to 1030 B.C.

Dr. Brady said this was the earliest radiocarbon date for cultural material in Honduras. Among the artifacts in the burial chamber were broken pieces of jade, 20 intact or restorable ceramic vessels and two large marble bowls.

Surveys of the mounds found less than a mile from the cave showed that the settlement stretched over an area more than a quarter of a mile wide, making it the largest occupation site known in Honduras from such an early time.

The buried traces of dwellings and their arrangement suggested that the villagers represented a well-organized and probably socially stratified population resembling other cultures in Mesoamerica, the region of Mexico and much of Central America.

Dr. David McJunkin, director of the University of Wisconsin's Radiocarbon Laboratory in Madison, reported that he was able to extract bone protein from two skeletal samples taken from the cave. An analysis of carbon isotope ratios indicated that neither person was subsisting on corn, he said.

Dr. Brady said, "We have an interesting situation in which the site combines a Mesoamerican architectural pattern with a lower Central American subsistence pattern."

If the people did not depend on corn, as was customary by this time in Mexico and most of upper Central America, then their staple was probably manioc, Dr. Brady said. These dietary differences could be further evidence that the indigenous people of Honduras represented, as early Spanish accounts indicated, a mixture of ethnic groups occupying borderlands between the prevailing cultures of lower Central America and the upper territory.

"It will require years of painstaking archaeological investigation and extensive scientific analysis to prove incontrovertibly that an advanced, complex civilization did flourish in the Mosquitia region of Honduras more than 3,000 years ago," said Dr. Brady.

—JOHN NOBLE WILFORD, January 1995

# Corn in the New World: A Relative Latecomer

A NEW TECHNIQUE for dating ancient organic matter has upset thinking about the origins of agriculture in the Americas. The earliest known cultivation of corn, it now seems, occurred much more recently than had been thought—4,700 years ago, not 7,000—and scientists are perplexed as they ponder the implications.

The new date means that people in the New World, in the Tehuac n Valley of the central Mexican state of Puebla in particular, probably did not begin growing their most important crop until as much as 4,000 or 5,000 years after the beginning of agriculture in the Old World. Hunter-gatherers who settled along the Jordan River Valley managed to domesticate wild progenitors of wheat and barley as early as 9,000 to 10,000 years ago, and thus became, as far as anyone knows, the first farmers anywhere. Perhaps such a lengthened time gap could suggest clues to the circumstances favoring the transition to agriculture, one of the foremost innovations in human culture.

The new evidence, said Dr. Gayle J. Fritz, a paleobotanist at Washington University in St. Louis, "makes it necessary to begin building new models for agricultural evolution in the New World."

But reliable as they may be, are the new ages definitive? Because all the ancient corn specimens examined so far were fully domesticated, scientists suspect they have yet to find the intermediate and earliest examples of cultivated corn. They may have been looking in the wrong places.

Dr. Lawrence Kaplan, a botanist at the University of Massachusetts in Boston and a specialist in dating ancient plants, cautioned that it was premature to revise the chronology of New World agriculture. "We ought to reserve judgment on whether the maize for Tehuac n is really as old as it's

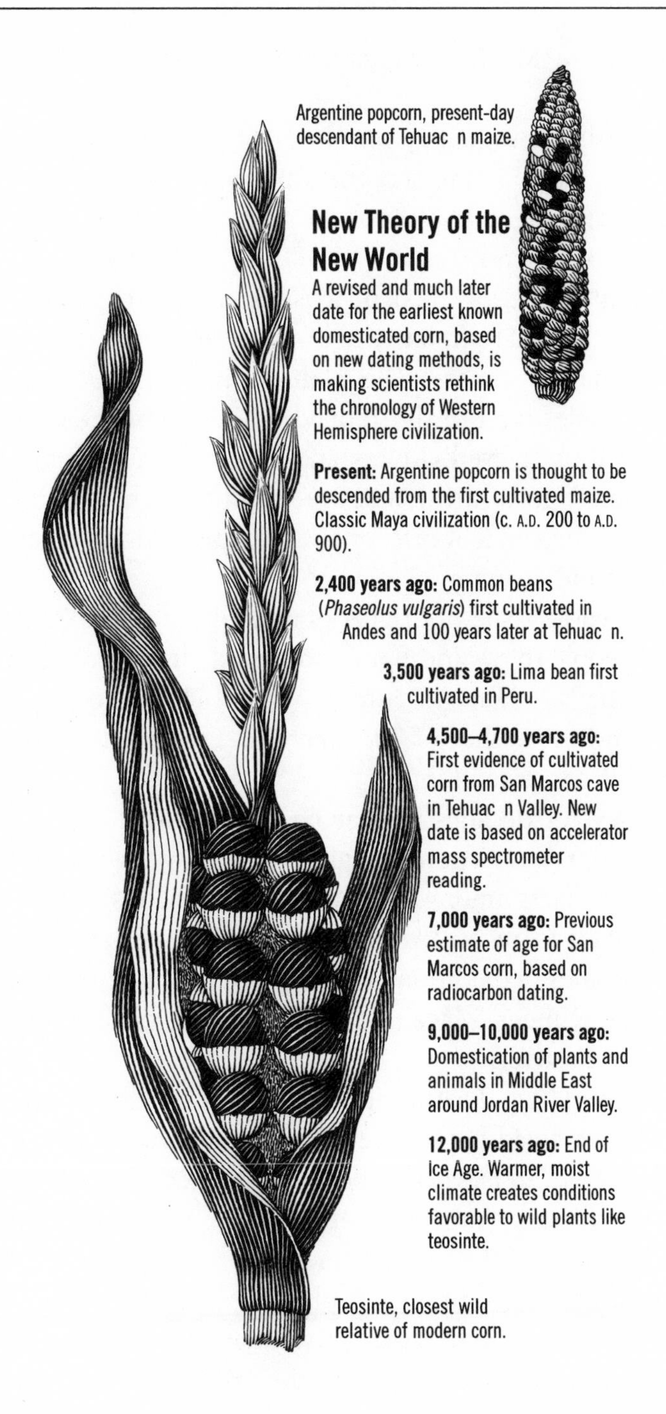


Brian Callanan

going to get in Mexico," he said. "Somewhere else, there may be older stuff."

Botanists are urging archaeologists to widen their search for evidence of early agriculture in Mexico, the only country where the nearest wild relatives of maize are native. Look in places where the wild teosinte grows, botanists recommend.

In many parts of Mexico, teosinte, an annual plant that shows the greatest biochemical similarity to domesticated corn, is still called *madre de ma z,* "mother of maize." The plant thrives in the verdant Balsas River basin, 150 miles west of the Tehuac n Valley, but the area has never been systematically surveyed. Rivers and lakes, moreover, are just the places where animals go to drink and are easy prey, where fish can supplement the diet and the soil is moist for planting, all conditions encouraging early settlements and farming.

"The whole issue of origins of agriculture in the Americas is still out there for people to try and figure out," said Dr. Bruce D. Smith, an archaeologist at the National Museum of Natural History of the Smithsonian Institution and the author of *The Emergence of Agriculture,* a book published by Scientific American Library.

For several decades, archaeological research in this field had been somewhat dormant. Archaeologists may have been discouraged by the paucity of artifacts among the remains of corncobs in the Tehuac n caves; nothing much to reconstruct the lives of the people who were the corn farmers. Besides, expeditions could count on more fruitful hunting in the ruins of the Olmecs, Maya and Aztecs, whose civilizations afforded more flamboyant discoveries.

So it was that the timing and pattern of early farming in the New World seemed fixed beyond serious questioning. Corn, or maize, known scientifically as *Zea mays,* had been established as the first American crop. It was the dietary staple in Mexico and eventually became the same throughout most of the two continents.

This decisive cultural step, planting and harvesting, was confidently dated at 7,000 years ago, based on standard radiocarbon analysis of material found in the Tehuac n caves in the 1960s by Dr. Richard S. MacNeish of the Andover Foundation for Archeological Research in Andover, Massachusetts. Buried in the dry sediments were two-inch-long ears of corn, each with eight rows of six to nine tiny popcorn-like kernels—a poor foretaste of sweet corn on the cob.

But one thing kept puzzling some scholars. These early New World farmers appeared to be seasonally mobile hunter-gatherers who visited the Tehuac n Valley just long enough to plant and harvest a crop, then moved on to where the hunting might be better. Indeed, it has long been a tenet of pre-Columbian anthropology that it was the domestication of corn, providing a steady source of food and thus increasing populations and encouraging a more sedentary life, that cleared the way for complex societies.

In the Old World, though, the sequence was reversed: sedentary life first, then agriculture. People there typically settled into communities near where wild animals and plants were abundant and then over time learned to increase and regularize their food supply through domestication of certain animals and plants, thus making the transition from hunting and gathering to agriculture. The more recent corn date, Dr. Fritz thinks, could have given the hunter-gatherers more time to experiment with possible sedentary living before taking up agriculture.

"I would not be at all surprised to find sedentary life before agriculture, probably in river and lake areas," she said. "But we don't have any evidence for it. We haven't really been looking."

In her own research in northeastern Louisiana, Dr. Fritz has already found evidence of other early American societies of hunter-gatherers leading sedentary village life. This way of life was practiced in many places in eastern North America in the centuries before Columbus. The search for evidence of this in Mesoamerica also stands high on the agenda of research into early American farming.

"The problem in Mexico is, the information we have on early sedentary villages is not very good," said Dr. T. Douglas Price, a University of Wisconsin archaeologist who specializes in the study of early agriculture. "We've been trying to encourage more people to investigate the archaeology of the early farmers."

Many of the old assumptions about New World agriculture are being reexamined in light of the new age estimates for early corn, reported in 1989 by Dr. Austin Long and colleagues at the University of Arizona in

Tucson. The scientists applied the new technology of accelerator mass spectrometry, which overcomes a serious limitation of conventional radiocarbon dating: the sample-size barrier.

In the standard method, developed in the 1940s, scientists could determine the age of once-living material, a piece of wood, cloth or corncob, by detecting and counting the decay rate of the radioactive isotope carbon-14 in the material. But this meant destroying a large sample to get the five grams of carbon necessary for the test. In the case of the early Mexican corn, the specimens were too small and too few to part with.

So archaeologists had done the next best thing. For the destructive radiocarbon tests, they used large samples of other organic material, usually charcoal found in the same sediments with the corncobs and kernels and thus assumed to be contemporary. It was an indirect measure and not very reliable, as they have found out. Seeds and other small objects have a way of being displaced in sediments, shifting up or down by the actions of burrowing animals, moisture and other disturbances.

With accelerator mass spectrometry, scientists can determine the age of samples as small as one thousandth the size of those required for the conventional method. Just a pinch of a cob or husk will do; not the whole thing. Rather than counting decay events, the particle accelerator separates and counts directly the carbon-14 atoms. This gives the time elapsed since the material was alive.

In addition to the corn dating, the technique has been used by archaeologists to date the skulls of horses found with chariot remains in burial mounds in Kazakhstan, leading them to conclude that the earliest known chariots came from this region 4,000 years ago. This is earlier than Russian scientists had estimated, and several centuries before the best evidence for chariots in the Middle East. Also, French and Spanish scientists recently used the technique to show that painted bison on the ceiling of the Altamira cave in northern Spain were painted not at the same time but centuries apart.

The earliest Mexican corn samples proved to be 4,700 years old; others were as recent as 1,600 years old. Dr. Kaplan has used the same technology to test the ages of primitive beans in Mexico and South America, once placed at 8,000 to 6,000 years ago. Like corn, domesticated beans, *Phaseolus vulgaris,* also turn out to be younger than thought—about 2,300 years at Tehuac n and 2,400 years in the Andes. Lima beans from Peru were dated at 3,500 years.

As the new findings undermined the record for a much earlier New World agriculture, Dr. Fritz grew impatient with textbooks and some professors who persisted in using the old corn dates and with researchers doing little to incorporate the new dates in their interpretations of early agriculture.

Writing in the journal *Current Anthropology* in June 1994, she urged colleagues to "forge ahead with drastic revisions for New World agricultural beginnings based on the earliest good dates available" rather than to cling to chronologies unsupported by solid evidence.

Several scientists, notably Dr. Dolores Piperno of the Smithsonian Institution Tropical Research Institute in Panama, insist that they have pollen and other evidence for domesticated corn and other plants in South America 7,000 years ago. But Dr. Fritz said she remained unconvinced by claims that corn farming had spread into Central America and northern South America before 5,500 years ago. Other botanists familiar with the work tended to side with Dr. Fritz.

Acknowledging that Dr. Fritz was correct to wake up scientists to the new data, Dr. Kaplan of the University of Massachusetts cautioned that the new dates for corn and beans "in no way represent the ultimate answer" concerning the fateful time at which early Americans turned to farming and cultivated a wild plant that became corn. After it was discovered by the rest of the world, corn became the third largest crop, after wheat and rice.

—JOHN NOBLE WILFORD, March 1995

# Tomb Find Suggests a Royal Family Murder

SOUTH OF CHICH N ITZ , the spectacular Maya ruins in the Yucat n Peninsula of Mexico familiar to archaeologists and tourists, lies the plain of Yaxuna. Dark mounds of stone rubble rise out of the fields of corn like islands in an archipelago. Excavations at these mounds have now revealed scenes out of the past that are both regal and macabre.

Archaeologists working at Yaxuna (pronounced YASH-oo-nah) discovered two tombs that over 16 centuries had somehow escaped the depredations of looters. These are the first undisturbed ancient tombs to be scientifically investigated in the Yucat n, and their contents are seen as striking evidence that the rule by kings so common in the Maya cities of the southern lowlands, in present-day Guatemala and Belize, extended to the northern lowlands—and so did a pattern of ruinous conquest warfare.

When archaeologists opened one tomb, a vaulted chamber inside an enormous pyramid, they found the skeleton of a man laid out in majestic repose. Near his head was a royal diadem jewel, and at the groin his hands clasped three round jade beads, the symbol of the celestial birthplace of the most important ancestral god in the Maya creation story. The man also wore shell earplugs that were the special mark of the god Chak. Surrounded and adorned by such regalia, this man, who must have been powerful in life, went into the Maya otherworld not only as a king but also as a god.

The second tomb, in a smaller pyramid nearby, held the remains of 12 to 15 people who appear to have been buried at the same time. At least three were women, one of whom was in an advanced stage of pregnancy. Of the six skeletons identified as men, one, by the shell and jade adornments and the stingray spines in his groin, was probably a king. One of the three children was buried with a large ceramic doll cradled in its arm. Another was buried with a fine royal pendant, indicating that this was an heir to the kingship.

After weeks of analysis of the artifacts found there, including shell pendants with the goggle-eyed figures associated with other Maya kings, archaeologists are almost certain that the second tomb was the end of the line for a defeated royal family, presumably captured and executed together.

"We may have a Romanov situation here," said Dr. David Freidel, leader of the discovery team, referring to the mass execution of the deposed Czar Nicholas II and his entire family by the Bolsheviks during the Russian Revolution. Dr. Freidel is a professor of anthropology at Southern Methodist University in Dallas.

One possible interpretation is that the mass burial attests to the capture and annihilation of a rival ruling family by the king of Yaxuna. But signs of the ritual destruction of the residential palace at Yaxuna, Dr. Freidel said, suggest instead the violent overthrow of the Yaxuna ruling family itself and the death of its members at the hands of a usurper king.

Indeed, the single skeleton in the first tomb could be that of the usurper, whose family would have continued governing Yaxuna. Pottery styles indicate that the two tombs were approximately contemporary, with the single-skeleton tomb being somewhat more recent. Radiocarbon dating of ashes in the tombs should provide more precise dates. A planned genetic analysis of DNA samples from bones in the second tomb should establish whether these individuals were members of the same family.

Whether the eventual interpretation will be that of a victorious or a vanquished Yaxuna king, Dr. Freidel said, the discoveries are "clear evidence that there were not only early kings in the Maya north, but that they practiced the methods of conquest warfare we are beginning to see in the southern lowland kingdoms of the classic period."

The classic period, the high point of Maya civilization, one of the most accomplished in pre-Columbian America, lasted from A.D. 200 to 900. The Yaxuna tombs have been tentatively dated at between A.D. 300 and 350. But the new research showed that Yaxuna had been a royal capital since the first century. Chich n Itz , only 10 miles away, did not rise to prominence until the eighth century.

Until recently, scholars had tended to idealize the Maya as a people of monumental architecture and unusually refined, peaceful ways. Excavations and the decipherment of royal inscriptions have in the past few years produced a considerably revised image of them as a not-so-peaceful people with rulers who engaged in rampant warfare and territorial expansion that probably contributed to the civilization's downfall. Kings of the large southern cities boasted of their triumphs in stone inscriptions and adopted ritual symbols and practices associated with war.

In their book, *Maya Cosmos*, published by William Morrow & Company, Dr. Freidel, Dr. Linda Schele of the University of Texas at Austin and Joy Parker, a writer, described Maya warfare as a sacred activity closely tied to the people's ritual life.

"All of the important rituals that took place in the great urban centers required the death of a sacrificial victim," the authors observed. "But the death of these valiant war captives was also a celebration of life born out of death, as well as the triumph of humanity over the dreadful diseases and misfortunes that threatened their existence."

Since royal hieroglyphics have yet to be uncovered in the northern lowlands, archaeologists lacked any literary basis for knowing if these people had similar forms of royal government. The Yaxuna discoveries enabled the scientists to correlate ritual and funerary symbols in the north with those from the south, where their meanings had already been determined with the help of associated inscriptions. The similarities were many, the archaeologists noted, and so must be the meanings.

Assisting Dr. Freidel in the excavations and analysis were Charles Suhler, a graduate student at Southern Methodist; Sharon Bennett, an independent forensic anthropologist; Traci Ardren, a graduate student at Yale University; Fernando Robles of the National Institute of Anthropology and History in Mexico City; and Rafael Cobos of the University of Yucat n in Merida. The research was supported by the Selz Foundation of New York.

For the archaeologists the most telling clues that Yaxuna had been a royal capital were the several greenstone diadem jewels, each no bigger than a quarter, uncovered in both tombs. They bear the face of the Sak-Hunal, meaning White Eternity, one of the most pervasive symbols of kings, with originals going back to the Olmec civilization preceding the Maya. Dr. Schele, an art historian, has dubbed this image the jester god because its characteristic headdress reminded her of the caps worn by court jesters in medieval Europe.

The single skeleton in the first tomb was laid out with his head oriented to the west, the direction of the setting sun. The Sak-Hunal jewel near his head had presumably once adorned a cotton headband. This declared the man's status to the gods and ancestors, and for Dr. Freidel it clinched the identification of him as a king.

Two of the ceramic pots in the tomb afforded a glimpse of Maya ritual practices. They are vessels decorated with pictures of people giving themselves enemas. From other art elsewhere, archaeologists have known that the Maya elite used enema potions, often fermented honey, to induce trances in order to commune with the gods. In this way, the intoxicants entered the bloodstream almost immediately and in full, unfiltered measure.

Dr. Freidel said this was the "first clear association of this practice with a king."

In the second tomb, Ms. Bennett noted evidence of another familiar Maya practice, genital bloodletting, which was also associated with kings and their communion with gods. The stingray spines found at the groin of one male skeleton had probably been inserted in his penis. Pictures studied elsewhere have showed gods dancing with sharp spines stuck in their penises. In Maya belief, scholars have inferred, genital blood was especially powerful and holy and thus favored in such rituals.

Ms. Bennett concluded that the 12 to 15 skeletons in the second tomb were buried at the same time because there is no sign of disturbances of the kind usually associated with periodic reopening of a chamber for additional burials.

Other artifacts in the ruins, including six goggle-eyed shell pendants and an owl jewel, led the archaeologists to conclude that the Yaxuna kings at this time, like those in the south, were engaged in conquest warfare. Both types of figures have been recognized as symbols of this increasingly destructive violence between rival kingdoms. Dr. Schele has called this Tlaloc-Venus warfare—Tlaloc because this god is prominent in the military regalia and ceremonies and Venus because the movements of this planet seemed to figure in the timing of war.

Excavations turned up several artifacts, a bone carving of a war bonnet and a tripod urn, that suggest the influence of Teotihuac n on the Yaxuna society. Teotihuac n was a large city in the central highlands near what is now Mexico City. From other evidence it is known this city exerted a widening influence through trade and other contacts in the Maya lowlands from the third through fifth centuries.

As they pondered the tombs of Yaxuna, Dr. Freidel and Mr. Suhler, who is writing his Ph.D. dissertation on the findings, began to conceive the story of what they think happened there.

Sometime in the early fourth century, the kings of Yaxuna adopted the rituals and practices of conquest warfare from Teotihuac n. Other kings in the Maya lowlands had done likewise. In one of the wars, an enemy defeated the king of Yaxuna and destroyed his residential palace stone by stone, crushing the plaster facade and burying it.

The lord chosen to replace the vanquished king of Yaxuna, probably a brother of the victorious king, destroyed one of the temples and had it rebuilt as his own. There he stored the losing king's jewels and, as part of his own accession to power, had the losing king and his entire family sacrificed in ritual fashion. They were entombed in the rebuilt temple. In Maya thinking, the victor, by having his own palace in the rebuilt temple with the loser's tomb, had transformed his enemies into his ancestors.

The new royal family ruled Yaxuna for some time, burying one of their kings, perhaps the usurper himself, in a pyramid of the north acropolis. From the grandeur of the pyramid, he must have wielded power over many towns and villages across the plain of Yaxuna, in a time of kings long before Chich n Itz  became an imperial center.

—JOHN NOBLE WILFORD, January 1994

# Volcano Captured Corn, Chilies and House Mice

IT WAS EARLY EVENING in the rainy season when the villagers felt the ground shake and then probably trembled themselves at the sight of blasts of steam in the distance. They dropped what they were doing and fled, abandoning their thatch-roofed huts, fruit trees in full ripeness and unwashed dinner dishes.

That was 14 centuries ago, around the year 590. The people saved their lives but lost their village, which was buried virtually intact under more than 16 feet of fine ash from an erupting volcano half a mile away. Only now are archaeologists digging through the ash, at the site they call Ceren in present-day El Salvador, and finding remains of the village so well preserved that it has been stamped with the inevitable sobriquet of Central American Pompeii.

Paleoethnobotanists, in particular, are having a field day. At Ceren, these specialists in the uses of plants by early people have uncovered stores of ceramic vessels filled with beans, squash, cacao and other plant foods—dried but still recognizable after all these centuries. They also gathered charred seeds, burned in the volcanic heat, which could be identified under the microscope.

As they examined the site in finer detail, botanists have even identified plants that had disappeared, though not without a trace. Some left delicate impressions of leaves, seeds or stems in the ash. Some that had completely decomposed left cavities in the packed ash. The cavities served as molds, which, when filled with dental plaster, produced clear replicas of plants. In this way, plaster casts re-created the corn cribs filled with unshucked ears as they looked the evening the volcano struck.

Archaeologists and botanists said Ceren is revealing for the first time much about the everyday lives of ordinary people in sixth-century Mesoamerica: their dwellings and crops, what they ate and how, their economy and how it contributed to the classic period of Maya civilization. Most explorations have concentrated on the grand ceremonial centers of the Maya rulers and priests.

"What we learn from Ceren," reported a team of scientists led by Dr. David L. Lentz of the New York Botanical Garden in the Bronx, "is that people living on the fringes may have made a contribution to the development of Mayan culture by producing surpluses and providing resources necessary to sustain the larger centers."

The director of Ceren excavations, Dr. Payson Sheets, an archaeologist at the University of Colorado at Boulder, said the stores of food, the abundance of finely painted ceramics and the evidence of trade goods showed this to be a prosperous community of farmers. In one of the more humble households, archaeologists counted more than 70 ceramic vessels.

"It's surprising, but also sad," Dr. Sheets said in an interview. "The standard of living fourteen hundred years ago was higher than it is now among the peasants of El Salvador."

Such archaeological discoveries at a farming community, important under any circumstance, are especially rare in tropical climates where moisture spells rapid decay for all organic materials, thatch, twine and the dinner beans, unless they happened to be sealed in the dry, airless depths of ash. The bean collection from Ceren is one of the largest found at any Mesoamerican archaeological site.

The existence of the buried village first came to light in 1976, but it was more than a decade before intensive excavations got under way. So far, 11 buildings have been uncovered: dwellings, an obsidian workshop, a food storehouse, a sauna for ritual sweat baths, a religious center and a community hall. The structures had adobe walls and roofs of thatch made from a type of grass that is now extinct, apparently killed off by alien grasses from the Old World. Found in the surviving thatch were the bones of mice that had infested the roofs.

Ground-penetrating radar surveys have detected the sites of many more dwellings, leading to estimates that 200 to 300 people lived in the village. Researchers are not sure if these were Mayans or members of local ethnic groups, either Lenca or Xinca. The nearest major center of Maya culture, then in its heyday, was Copan, about 60 miles away in Honduras.

Whoever they were, these people had made this site in a fertile valley their home for several centuries. Archaeologists can almost pinpoint the time it all came to an end.

It was the rainy season of July or August, because evidence of fully ripe fruits of that time of year was found in abundance. It must have been early evening, probably between dinner and sleep. The men had certainly returned from work in the cornfields, because their tools were stacked near the dwellings. Food was still in pots on the hearths. Some serving bowls were empty but unwashed, still bearing the marks of people eating a semisolid substance, possibly cornmeal gruel, with their fingers. And it was before they had retired for the night, or else their sleeping mats would have been rolled out.

In an interview and a report in the journal *Latin American Antiquity,* Dr. Lentz said that an analysis of the plant remains showed that Ceren had developed a diverse agriculture of orchards, household gardens and large cornfields. At least at this time of year, they had a variety of foods to eat: corn, several kinds of beans, squash, chili peppers, avocados, nuts, cherries and other fruits. They drink cacao, the favorite beverage of the Maya. They got their animal proteins from deer, ducks, dogs and freshwater mollusks.

Dr. Lentz and his colleagues concluded that the Ceren farmers "had at least as varied a diet as the Maya nobility" living at Copan. Most of these foods would be familiar to Central Americans today. But Dr. Lentz said, "We had no record of chilies anywhere in the Mayan realm—here we find thousands of them, just everywhere, hanging from the rafters."

Cotton appeared to be a major crop, he said, with the fibers used not only for the villagers' own clothing but for trade. They may have used cotton cloth as exchange material for obsidian, exotic ceramics and other imported goods found at the site.

The botanists also found evidence that the people were grinding cotton seeds to extract the oil, possibly for cooking. Although it has never been demonstrated that the Maya knew about cottonseed oil, Dr. Lentz said, they probably did use it for cooking because they had few other sources of oil. Their only domesticated animals were dogs, turkeys and ducks.

Dr. Lentz, who also directs ethnobotanical research at the City University of New York, was joined in his work on Ceren by Dr. Marilyn P. Beaudry-Corbett, a ceramicist at the University of California at Los Angeles; Dr. Maria Luisa Reyna de Aguilar, a botanist in San Salvador; and Dr. Lawrence Kaplan, a biologist at the University of Massachusetts at Boston.

"Ceren provides part of the answer to the question of how the Maya were able to develop and maintain such large ceremonial centers," Dr. Lentz said. "These farmers were producing the agricultural surpluses that probably contributed to the civilization's wealth and culture."

—JOHN NOBLE WILFORD, April 1997

# An Ancient "Lost City" Is Uncovered in Mexico

IN A LUSH RIVER DELTA on Mexico's Gulf Coast, archaeologists have found temple mounds, ball courts and other traces of a sprawling pre-Columbian seaport city that flourished more than 1,500 years ago and may have been a vital center of ancient culture and coastal commerce.

A preliminary survey of the site, about 60 miles northwest of the modern city of Veracruz, has revealed the ruins of more than 100 earth-and-stone pyramids and other structures, some reaching heights of 130 feet, that had long remained largely hidden under dense vegetation. The core city and its suburbs extended over 40 square miles and were occupied by thousands of people, possibly more than 20,000—large for that time and region.

No one is prepared to say who these people were. The city, which existed between A.D. 100 and 600, rose after the disappearance of the Olmec civilization, once strong along the Gulf Coast, and centuries before the Aztecs of central Mexico. It was contemporary with the classic period of the Maya, but they lived several hundred miles to the southeast. It probably had strong cultural and trade ties with Teotihuac n, the powerful urban center near present-day Mexico City.

In any event, the ancient city, called El Pital for a nearby village, is thought by its discoverer to be one of the most important archaeological discoveries in the Veracruz region in more than 200 years.

The discovery announced in Mexico City and Washington was described in *National Geographic Research and Exploration*, a quarterly journal of the National Geographic Society. The society and the Selz Foundation of New York City helped finance the research by S. Jeffrey K. Wilkerson, an independent archaeologist who made the discovery.

In the journal article, Mr. Wilkerson said that El Pital "may well alter our concept of Mesoamerican culture history," calling the city "pivotal—in both time and space—to the emergence of classic civilization," the period of urban growth and cultural splendor that ran from A.D. 250 to 900 for many cultures in Mexico and Central America.

Mr. Wilkerson is an American who has lived in Veracruz for more than 20 years and conducts archaeological research through his own Institute for Cultural Ecology of the Tropics. He is also associated with the Smithsonian Institution's National Museum of Natural History. His explorations at El Pital were authorized by the National Institute of Anthropology and History in Mexico City.

"It's a very interesting site," said George Stuart, chairman of research of archaeological projects at the National Geographic Society. "It's important and needs to be investigated."

Like other archaeologists, however, Mr. Stuart cautioned that no systematic excavations have been conducted and until they are, nothing definitive can be said about the city's role in pre-Columbian culture.

But the prospect of finding elaborate ruins, possibly of a culture unknown until now, is exciting for archaeologists specializing in pre-Columbian exploration. The coastal regions north of Veracruz have been largely neglected because reconnaissance and excavations are difficult in the area's dense jungle. Also, research has long seemed to be more rewarding in the central highlands around Mexico City and in the Maya country to the south.

The El Pital site lies nine miles inland from the Gulf of Mexico, upstream on the Nautla River at the head of navigation. Mr. Wilkerson noted that in El Pital, unlike most Veracruz urban centers of that period, which were in defensible valleys and ridges, security "is likely to have rested on its direct governance of a broad region, far greater centralization than its immediate neighbors and alliances of lineage or commerce that place it at the hub of a regionally valuable network."

One of the best-known ancient cities in Veracruz was Matacapan, which sprang up in the south around A.D. 400. Archaeologists have surmised that this city thrived by long-distance trade and was dominated by possibly colonial ties to Teotihuac n.

In his survey of El Pital, Mr. Wilkerson said he found many Teotihuac n-style ceramics. But local styles, particularly in figurines and vessels, were also strong, he said, indicating that El Pital "is likely to have had a far more complicated role than that of a trade way station or Teotihuac n outpost."

Mr. Wilkerson noted that some murals in Teotihuac n depict a riverside scene of raised agricultural fields, a farming practice at El Pital, and dense tropical flora. Since nothing like that could be found in the semiarid plain of Teotihuac n, he said, this could be an image of El Pital "as a sort of Eden," reflecting its apparent role as a major center for food production.

El Pital is significant, Mr. Wilkerson said, "because it appears to be the principal end point of an ancient cultural corridor that linked the north-central Gulf Coast with the cities of central Mexico." One research goal will be to determine the city's importance as a seaport and the extent of its coastal reach in trade. Some scholars have even suggested that corn and some cultural practices traveled from central Mexico to the Mississippi River Valley about this time, by either overland or sea trade.

Dr. Norman Hammond, a Boston University archaeologist who specializes in Mesoamerican studies, said it was impossible to conclude from present knowledge whether Mexico was the direct source of these innovations among the Indians of the Mississippi Valley.

The newly discovered site resembles in at least one respect another ancient city in the region, El Tajin, found some 40 miles away in 1785. A prominent feature in both were courts used in a ritual ballgame that often involved sacrificial decapitation of some players. Fragments of stone depicting aspects of the games, including figurines representing perhaps sacrificed ball players, were found at the site.

For centuries the El Pital site was obscured by rain forest. The land was opened to agriculture in the 1930s and is now heavily planted in bananas and oranges. People who lived there and worked the fruit plantations took the mounds for granted, assuming they were natural hills.

"This reminds us," Mr. Wilkerson said, "that the time has come in the largely deforested tropics to carefully search for the 'lost cities' we have overlooked."

—John Noble Wilford, February 1994

# Lost Civilization Yields Its Riches as Thieves Fall Out

A FALLING-OUT among thieves and a trail of looted treasure led in 1987 to one of the most spectacular archaeological discoveries of this century: the glittering royal tombs of the Moche civilization, which dominated northern Peru from A.D. 100 to 800, centuries before the rise of the Inca Empire.

From time to time, the villagers of Sipan had supplemented their incomes plundering ancient cemeteries and adobe pyramids of the Moche (pronounced MOH-chay). One night, while digging deep into an eroded pyramid, they broke into one of the richest funerary chambers ever looted. They filled gunnysacks with decorated ceramic vessels and gold and silver objects. After an argument over the division of the loot, one of the dissatisfied thieves went to the police, who confiscated the material and learned of its source.

Archaeologists could hardly believe their good fortune. Three spacious burial chambers in the pyramid held an astonishing array of gold and silver artifacts, elaborate headdresses, painted vases rivaling the art of ancient Athens and ceramics decorated with scenes of hunting, combat and ritual. One of the tombs was the resting place of the most important Moche ruler, the Warrior Priest, who has come to be known as the Lord of Sipan.

These were the richest pre-Columbian tombs ever excavated, scholars say. In their even more enthusiastic moments, archaeologists rank the discovery right up there with such famous coups of the century as Tutankhamen's tomb, the Sumerian tablets from the tombs of Ur, the Dead Sea Scrolls and the terra-cotta statues of the Qin Dynasty warriors uncovered near Xian, China.

Dr. Walter Alva, chief archaeologist at Sipan and director of the Bruning Archaeological Museum in Lambayeque, Peru, accompanied the police in 1987 on the predawn raid of a suspected looter's home that broke open the case; one of the illegal diggers was killed in gunfire. Over the next two years, while armed guards kept watch, he directed Peruvian excavations of the three tombs, finding artifacts unlike any previously seen and gaining insights into Moche art and society.

"We have learned more about the Moche from these tombs than from all the art objects dug up in centuries of looting," he said.

Moche art is invaluable to scholars of the culture because the people had no written language. Their artisans had to speak for the culture, and they left the only stories of daily life, hunting, war and religious ceremony in the scenes on ceramics and textiles.

"They almost made up for the lack of a written language with their pottery with its incredible iconography," Dr. Craig Morris, a specialist in pre-Columbian Andean culture at the American Museum of Natural History, said.

The Moche potters also had a flair for lifelike portraiture in clay, surpassing anything produced by other pre-Columbian American cultures. Dr. Christopher B. Donnan, director of the Fowler Museum of Cultural History at the University of Calfornia, Los Angeles, was surprised to find that important individuals he had identified in Moche artworks were not mythological figures, as he had suspected.

"What is depicted artistically often has a basis in fact," Dr. Donnan said at the opening of an exhibition at the Fowler Museum of some of the most impressive material from the tombs. "We now realize that the art is in fact showing real people, and what we suspected were supernatural activities were real events that occurred in Moche life."

The ornaments, scepters and other regalia associated with the Warrior Priest in previously discovered art, he noted, were found in the tomb around the coffin; the priest in art was real, and this was his tomb. He died near the end of the third century. In other tombs, other figures from art were recognized as individuals who had specific religious roles.

The Moche people flourished along a 250-mile stretch of Peru's northern coast for seven centuries, living in the fertile valleys of rivers that flow out of the Andes and developing elaborate irrigation systems for farming. In each valley rose a large ceremonial center with palaces and pyramids, one covering almost as much ground as the Great Pyramid at Giza in Egypt. Some of the royal complexes were surrounded by settlements of up to 10,000 people. The Moche were contemporaries of the Nazca in southern Peru, the people associated with the mysterious markings, or lines, in the desert plateau.

Among the artifacts taken from the earliest tomb found, that of an individual called the Old Lord of Sipan, are gold and gilded-copper masks, ear and nose ornaments and a necklace of 10 intricately carved gold beads, each depicting a spider with a body in the form of a human head. Anthropomorphized spiders occur frequently in the tombs and are seen as analogs to warriors in Moche culture; spiders capture their prey, tie them with ropes of web and later extract their vital fluids, just as Moche warriors captured the enemy, tied them with ropes and later drank their blood.

Among the artifacts found in the Warrior Priest's tomb are curious ornaments known as backflaps, which were apparently worn by warriors. They were made of gold or silver and were suspended from the back of the warrior's belt with a large, flaring edge hanging down. The upper part was decorated with the figure of a deity with a large fanged mouth.

The Moche were especially skilled metalworkers, using gold, silver and copper in innovative combinations and developing a chemical plating process for gilding copper objects. This technology had disappeared among the societies encountered by Europeans after 1492.

The Moche civilization collapsed near the end of the eighth century for reasons unknown. When the Spanish arrived in Peru in 1528, the Incas ruled the area where the Moche had once flourished. But archaeologists have come to realize that much of the Inca art and technology was based on the innovations of earlier cultures, notably the Moche, who are known to posterity almost entirely through the beautiful ornaments and art from their tombs.

—JOHN NOBLE WILFORD, July 1994

## Evidence Shows Link of Two Centuries of Drought to Maya Decline

New evidence from a Mexican lake bed has revealed that an unusually severe drought about 1,200 years ago may well have contributed to the abrupt decline of classic Maya civilization.

Scientists from the University of Florida said an analysis of sediments beneath Lake Chichancanab on the Yucat n Peninsula provided "the first unambiguous evidence" for a period of extreme aridity between A.D. 800 and 1000. The drought was the harshest in the region in 8,000 years and coincided with the widespread collapse of Maya culture, a time of spreading warfare and finally the abandonment of many great cities of monumental architecture.

Although in recent years archaeologists have made many new discoveries in Maya ruins and managed to decipher telling hieroglyphic texts, the mystery of the civilization's sudden decline a thousand years ago persists. Various theories have implicated overpopulation, environmental degradation and bitter intercity rivalries leading to destructive warfare throughout much of the Maya lowlands of what is now Mexico, Guatemala, Belize and Honduras.

And now, in a report published in the journal *Nature,* scientists have added regional drought as a possibly critical factor in the civilization's fate. The findings from a chemical analysis of shell carbonate and gypsum, burned grasses and root fragments in cores from the lake bed might not alone support the drought theory, they said, but other geological evidence of low lake levels elsewhere in Mexico and Costa Rica suggested that the climate shift was just in Central America.

The study was conducted by Dr. David A. Hodell and Dr. Jason H. Curtis, both geologists at the University of Florida, and Dr. Mark Brenner of the university's department of fisheries and aquatic sciences.

In an accompanying commentary, Dr. Jeremy A. Sabloff, director of the University of Pennsylvania Museum of Archaeology and Anthropology and a Maya specialist, said the new findings added "another important component to the model that archaeologists are currently building to help explain the upheavals of the late eighth century and early ninth century A.D. in the Maya lowlands."

But the drought lasting at least two centuries was probably not solely responsible. The region, Dr. Sabloff noted, was "already under heavy stress" from political, economic and environmental problems.

"To a civilization facing a number of stresses both internal and external, the scarcity of water could have greatly increased the vulnerability of numerous classic Maya cities," Dr. Sabloff wrote. "The arrival of a drier regime would have exacerbated the perilous situation, ultimately causing the demise of elite power, the abandonment of many urban centers and important demographic and economic shifts from the southern to northern lowlands."

Such research into the causes of the Maya collapse could have implications for other cultures, ancient and contemporary.

By identifying the circumstances contributing to Maya successes and failures, Dr. Sabloff said, anthropologists might learn "how severe internal stresses in a civilization have to become before relatively minor climate shifts can trigger widespread cultural collapse."

—JOHN NOBLE WILFORD, June 1995

# Archaeologists Wonder at a City That Survived the Maya Collapse

ARCHAEOLOGISTS STRUGGLING TO UNDERSTAND how and why the classic Maya civilization collapsed so catastrophically in the ninth century A.D. keep turning up evidence that complicates their task. Among the ruins of grand cities suddenly abandoned they are finding some others in the Central American lowlands that not only survived but prospered long after.

Such a place was Xunantunich (pronounced Shoo-NAN-too-NEECH). New discoveries, announced by the University of California at Los Angeles, show that this classic Maya city survived the spreading chaos and remained a bustling center of art, religion and culture for another 150 to 200 years after what had until recently been considered a virtually total decline.

Scholars of Maya history said the findings supported the growing impression, based on several other recent excavations, that the civilization's demise was not as uniform as had been thought. The scholars so far are at a loss to explain why some cities were spared.

"This site compels us to revise archaeological thinking about the decline of the classic Maya and forcefully reminds us that we must stop talking about the Maya collapse during the eight hundreds as all-encompassing," said Dr. Richard M. Leventhal, director of the U.C.L.A. Institute of Archaeology and leader of the Xunantunich excavations.

As archaeologists have come to realize, the heyday of the Maya culture from Mexico's Yucat n south to Honduras extended from the year 250 to 900. In this classic period, the Maya built large cities with imposing pyramids, developed a complex writing system, mastered mathematics, compiled accurate calendars and engaged in widespread trade and efficient agricultural practices. They may not have been as peaceful as once imagined, and rampant warfare among rival city-states may have contributed to their undoing.

Some cities eventually recovered from the crisis, notably places in the Yucat n like Chich n Itz  and Uxmal, and flourished for a few centuries more. Xunantunich, Dr. Leventhal said, was a rare example in that it survived as "a bastion of social stability while classic Maya culture nearby was eroding."

The ruins of Xunantunich in Belize, which lie 70 miles west of Belize City, were discovered in the nineteenth century and are a popular tourist attraction. In new research this year, the U.C.L.A. team analyzed ceramics found in abundance at the site and determined that they were made in the tenth century, as much as a century after neighboring cities had fallen. The style of the ceramics, with distinctive piecrust rims, was similar to that of objects from the late classic period.

"These ceramics were clearly produced in a complex, functioning Maya city," Dr. Leventhal concluded.

Other evidence of the city's continuing vitality included a spectacular plaster frieze found on the west side of the Castillo, a 130-foot-high, pyramid-like temple. The frieze, more than nine feet high and 30 feet long, features a three-dimensional figure of a Maya ruler, ancestor gods, shells, earth monsters and dancing figures. These are images that in the Maya worldview symbolically linked Maya sovereigns with supernatural authority.

Researchers determined that the frieze was sculpted in the ninth century, the very time other, more dominant cities like Tikal, Seibal, Dos Pilas and Caracol were declining or had already been abandoned. At that time, Xunantunich was a thriving city of at least 10,000 inhabitants.

The dating of the frieze, Dr. Leventhal said, "tells us that not only was this work constructed relatively late in Maya history but that the city was still capable of undertaking large architectural projects even while once-powerful neighboring cities were falling apart."

Other archaeologists said the findings were an important and fascinating contribution to Maya research.

"This adds another brushstroke to the complex picture of Maya history," said Dr. Arthur A. Demarest, an archaeologist at Vanderbilt University in Nashville. Dr. David Freidel, a Maya specialist at Southern Methodist

University in Dallas, said, "The dating based on ceramics is pretty solid, and the logic of Leventhal's argument is compelling."

But Dr. Demarest, Dr. Freidel and others said they were not surprised, since excavations elsewhere as early as the 1970s had already caused archaeologists to realize that the Maya collapse was not as complete as once thought. Dr. Demarest emphasized that the findings should not lead to any major revision of thinking about the Maya decline. The fall of the Roman Empire, he noted, did not occur in one fell swoop.

Whatever happened at Xunantunich and a few other last-gasp places, Dr. Demarest said, "it's very clear that at the time Maya cultures were undergoing drastic changes across the whole area," adding: "There was a dramatic decrease in sociopolitical complexity, radical change in the cultural inventory. If this was not a collapse, I ask, what is?"

For most archaeologists, the question now is, what enabled some cities to survive the surrounding disaster? Some possible answers have been proposed, but as yet without much conviction.

Dr. Leventhal suggested that somehow there might have been some safety in Xunantunich's small size, compared to the much larger neighbors like Tikal, which had a population of 50,000. Xunantunich was founded relatively late in the classic period, around 700, probably as an offshoot of Naranjo, a city that died out in 830; Naranjo was in what is now Guatemala.

Xunantunich's position as a second-level city perhaps enabled it to keep out of the escalating warfare, as a kind of neutral Switzerland amid belligerent neighbors. Dr. Freidel said this hypothesis was attractive. Rulers of the larger cities, for defensive or expansive reasons, were always seeking alliances with other powerful rulers. Rulers of a secondary city like Xunantunich would have more modest ambitions and, if shrewd, might have been able to stay out of entangling alliances, Dr. Freidel suggested, and thus "not experience the epidemic chaos of the many wars."

Lending support to this thinking, Dr. Leventhal noted that there was no sign that any defensive walls had been erected around Xunantunich at the time other city-states were engaged in bitter warfare. Hastily built walls and other evidence of siege warfare were uncovered by Dr. Demarest in excavations at Dos Pilas in northern Guatemala.

Dr. David Pendergast, curator of New World archaeology at the Royal Ontario Museum in Toronto, was the first to find clues of lowland cities' somehow surviving the collapse. In the 1970s, his excavations showed that Lamanai, a classic city in northern Belize, kept going well beyond the time of the collapse. Since the few known survivor cities are in Belize, archaeologists wonder if that could suggest an explanation.

"By 1980," Dr. Pendergast said, "we had ample evidence of a continuum at Lamanai, and that the collapse was most pervasive in the interior lowlands. Lamanai was on a major river, an avenue out to the sea, and so could benefit more readily by trade with northern Yucat n and had access to wider food resources. It's highly likely that this was a factor in survival."

The same could be said for Xunantunich, archaeologists noted, because it was on a river and close enough to the sea to reach out and perhaps save itself through interregional trade and more extensive food sources.

Continuing explorations at Xunantunich, Dr. Wendy Ashmore, an archaeologist at the University of Pennsylvania, is to conduct a detailed survey of ruins revealing living patterns in the region surrounding the city. Dr. Leventhal plans to begin tunneling into the Castillo to look for more evidence of what it was like to be a surviving city in a dying civilization. Eventually, he said, he would like to find out why Xunantunich itself finally fell, sometime after the year 1000.

—JOHN NOBLE WILFORD, October 1993

# Move Over, Iceman! New Star from the Andes

ON AN ICY, 20,700-foot summit of the Peruvian Andes, archaeologists have found the well-preserved frozen remains of a young woman who apparently had been sacrificed to the Inca gods about 500 years ago. Her body was wrapped in finely woven wool, and she was wearing an elaborate feather headdress. Around her were rare ceramics and statuettes, artifacts of the religion that took her life on the sacred mountain.

Two more bodies were discovered at a slightly lower elevation of Mount Ampato in southern Peru. One of those was also female and partly frozen. Little more than the skeleton remained of the third body, probably a male.

From these haunting scenes of ritual death, scientists expect to learn much more about life among the Inca, whose empire spanned most of the Andes and the western coast of South America at the time of the Spanish conquest in the early sixteenth century. Preserved organs, tissues and fluids in the bodies could yield unfragmented DNA for genetic studies, as well as insights into Inca health and nutrition. The artifacts should provide valuable information about the still somewhat mysterious Inca religion.

On a more unscientific level, the discovery evoked the past not in stone or bone but in an eerie bodily form. In 1991, a hiker from the early Copper Age 5,000 years ago in Europe seemed to step into the present, with the discovery in the Alps of the frozen mummified corpse on the Austrian-Italian border. If the Alps could have its Iceman, as the find is popularly called, then the Andes now has its Icewoman.

The new discoveries were announced by Dr. Johan Reinhard, an American archaeologist and mountaineer. He described the findings in a telephone interview from Arequipa, Peru, and at a news conference there at the Catholic University of Santa Maria, where the bodies are stored in a freezer.

"It is certainly one of the most important discoveries in Peru since the Lords of Sipan," said Dr. Sonia Guillen, a Peruvian bioanthropologist who specializes in mummy studies. She was referring to the excavation in 1987 of a royal tomb at Sipan, in northern Peru, that contained a wealth of gold from a little-known pre-Inca civilization called the Moche.

Dr. Guillen, who is associated with Mallque, a biological research institute in Ilo, Peru, is to take a leading role in the conservation and study of the mummies. Dr. Konrad Spindler, an archaeologist at the University of Innsbruck in Austria and director of research on the Alpine Iceman, has arrived at Arequipa to help plan research on the bodies. The analysis will be financed in part by the National Geographic Society, which issued an announcement of the find in Washington.

"These bodies apparently are partly frozen but not freeze-dried—a first in the Andes," said Dr. George Stuart, chairman of the National Geographic Society's research and exploration, explaining the unusual nature of the bodies' preservation and their potential for scientific study. This will make it easier to study the thawed tissues.

Dr. Craig Morris, a curator of South American archaeology at the American Museum of Natural History in New York City, said the most significant part of the find might turn out to be the artifacts associated with the ritual sacrifices. The Inca, he said, worshipped the landscape and particularly the high mountains, which they deemed sacred and which they believed must be appeased with human sacrifices. Mountains to them were the source of water and weather and terror in the form of avalanches and blizzards.

What is learned about human sacrifices by the Inca could be compared to and contrasted with what is known about those of other early cultures.

Dr. Reinhard said that sacrificial victims were often prepubescent boys or young women, probably virgins, whose innocence would please the Inca deities. Being selected was considered a high honor. After considerable indoctrination, the victims presumably climbed the mountain and voluntarily submitted to their death at the hands of priests. They might be killed by a blow to the head, by strangulation or by smothering. They were buried under the snow and ice, where their bodies froze quickly. The bodies atop Mount Ampato had apparently remained frozen until now and had been buried until recent volcanic eruptions caused a collapse of the ritual platform at the summit and a shifting of ice at the burial site.

Dr. Reinhard is a specialist in cultural anthropology affiliated with the Field Museum of Natural History in Chicago and the Mountain Institute in Franklin, West Virginia. He has been studying the archaeology of the high Andes for more than 15 years, and knew that Mount Ampato was one of the summits sacred to the Inca.

In September 1995, he set out with Miguel Zarate, a Peruvian mountaineer, to climb to the top of Ampato. "It was free of snow," Dr. Reinhard recalled in an interview. "We saw feathers sticking out from the slope. We knew immediately there had been sacrificial offerings there."

The first body, wrapped in textile bundles, was lying in the open. The face had been exposed for some time and was badly damaged, Dr. Reinhard said. Looking around, he counted at least a dozen statuettes made of gold, silver, bronze or spondylous shell, a large oyster-like shell. They ranged in height from two and a half to six and a half inches. Some of the human figures, also offerings to the gods, were wearing perfectly preserved textile clothing and feather headdresses.

The two men also found the remains of the site where people had camped before making the final ascent to the top for the sacrifice ceremonies. They identified pottery, ropes, sandals, the remains of tent supports and even matted grasses that had probably been used to insulate their quarters against the cold winds.

"We have the whole context in which the sacrifices were conducted," Dr. Reinhard said.

It was a struggle for him to carry the frozen body down the mountain on his back, while Mr. Zarate cut steps in the steep ice. Then it was a 13-hour trek to the town of Cabanaconde, with the body on the back of a burro. The last lap was a bus ride from there to Arequipa, near the Peruvian coast. It is unclear how much of the body thawed during that interval.

The two men returned to the summit and found the other two specimens. The exact weight and measurements of the bodies cannot be determined until they are unwrapped. All three bodies are expected to be unwrapped and ready for study in a few weeks. But X rays of the first body have already determined that its organs are intact.

—JOHN NOBLE WILFORD, October 1995

# Mummies May Be of Incan Elite, After Conquest of "Cloud People"

An examination of more than 200 well-preserved mummies that were discovered in Peru shows evidence that these were elite members of an Amazonian society in transition after the Incans conquered the region 500 years ago.

Dr. Sonia Guillen, an archaeologist leading the research, said in an interview that most of the mummies were apparently Incan, but that they were surrounded by embroidered textiles, wooden idols, decorated gourds, pottery and other artifacts clearly belonging to the remote Chachapoya culture, called the "cloud people," that had only recently been overrun by the expanding Inca Empire.

Also found among the mummies were knotted strings of cotton or wool, known as *quipus,* which the Incans used for tallying populations in conquered areas and for other accounting and inventory purposes. By the various thicknesses and colors, the strings indicated numbers and other data, though scholars have yet to decipher the system.

"Here you see evidence of what cultural integration means," Dr. Guillen said. "This is what empires do."

Dr. Guillen, a Peruvian on the staff of the Bioanthropology Foundation, an international organization based in Britain that studies ancient human remains, said that investigation of the bodies and their furnishings over the next five years should yield insights into the interactions between the Inca and Chachapoya cultures. Only five of the 200 recovered mummies have been examined in detail, and 12 cave burial sites of more bodies have yet to be fully explored.

The discoveries were made in northern Peru at a place called Laguna de los Condores, in mountains near the headwaters of the Amazon. The area, west of the modern town of Leymebamba, is almost abandoned now, but was thickly settled in the Chachapoya heyday. These people flourished more than 1,000 years ago, long before the Incas rose to power, and dominated the Amazon headwaters until their fall in 1470, followed by the Spanish conquest in the early sixteenth century.

The mummies were found in six natural mausoleums set in the cliff some 300 feet above the lake. Considering the humidity of the cloud-shrouded forest, it was remarkable that any of the bodies survived, but they had been eviscerated, carefully mummified and wrapped in embroidered bundles. Most remained fully fleshed.

"This is a miracle of preservation," Dr. Guillen said, noting that most of the ancient mummies found in Peru—and there have been hundreds in recent years—came from the extremely dry desert in the south.

She said the care with which the bodies were prepared, as well as the quality of the grave goods, suggested that this "was an important group of the population, an elite."

Looters had cut open the cotton and wool wrappings of some of the mummies and probably stolen necklaces, pottery and other materials. The National Culture Institute in Lima authorized Dr. Guillen to oversee the immediate rescue excavation of the site. She was accompanied by a film crew for the Discovery Channel.

The National Geographic Society announced that the skeletal remains of another ancient Peruvian human sacrifice had been recovered near the icy summit of Mount Ampato, where the frozen and naturally mummified "Ice Maiden" was discovered two years ago.

The new mummy, thought to be a young female as well, was found wrapped in Inca textiles and accompanied by plates, pots and a female figurine, made of shell, similar to the items found with the Ice Maiden. Archaeologists concluded that maiden was taken up the mountain as part of an Inca ritual offering to the gods about 500 years ago. And from all appearances, the new discovery tells the tale of a similar human sacrifice.

One difference is that the new find is little more than a skeleton. The flesh may have burned away in a lightning strike or decomposed when the surrounding soil thawed. The first mummy was frozen with flesh and organs intact.

Both of these discoveries were made by Johan Reinhard, a mountain explorer with the geographic society. On the latest expedition he was accompanied by Jos  Antonio Ch  vez, an archaeologist at Catholic University in Arequipa, Peru.

—John Noble Wilford, December 1997

# Using Earth's Magnetic Pull to Track Ancient Treasure

THERE WOULD NOT BE another miracle here today.

In the Mexican village of Cruz de Milagro, named for a cross that legend said was miraculously dropped here by an angel in 1905, an American geophysicist, Sheldon Breiner, stood at the lip of a gaping eight-foot-deep hole.

Holding the world's most sensitive portable magnetometer, a device that detects minute variations in the earth's magnetic field, Dr. Breiner had hoped to find an ancient artifact buried at this spot, maybe a legacy of the Olmecs, one of America's first and most mysterious civilizations. After all, a stone figure of an Olmec prince was discovered by villagers while they were excavating for a water tank in 1961.

However, using a magnetometer as an archaeological tool to "see" human history buried beneath the ground is still as much an art as it is a science.

Three feet below the surface was indisputable evidence of a civilization: A decidedly twentieth-century metal pipe protruded from the wall of the dig, a mute sign that even the most accurate magnetometer can only give the most general hint about what lies beneath the surface.

Surrounded by about 20 local children and villagers and a handful of diggers hired by the archaeological team, Dr. Breiner's immediate concern seemed more the disappointment this dry hole might cause the local community than the sterile red earth the diggers had encountered six feet below the surface.

"The archaeologists teased me and told me that I better leave town in a hurry if we didn't find anything," he laughed, looking around at the crowd who appeared to be more curious than angry.

Dr. Breiner's track record is usually much better. A pioneer in the use of magnetometers in archaeology, mineral exploration, earthquake detection, weapon detection, treasure hunting and even military applications, he has located during the past three decades two of the 17 so-called "colossal" Olmec heads that had been found so far.

The striking eight-foot- to 15-foot-high basalt figures, some weighing as much as 14 tons, are the most remarkable artifacts of the Olmecs, an advanced civilization that flourished as early as 1500 B.C. in the humid, fertile coastal lowlands of the Mexican states of Veracruz and Tabasco. The Olmecs quarried basalt in the Tuxtla Mountains and floated the stones to their villages along the rivers that flowed to the Gulf of Mexico. The civilization mysteriously vanished around 400 B.C., centuries before the rise of the better-known Mayans who greeted the Spanish conquistadors in the sixteenth century.

It had been 30 years since Dr. Breiner first joined an archaeological expedition to explore an Olmec site, one near San Lorenzo Tenochtitl n where, in the late 1940s, Matthew Stirling of the Smithsonian Institution had discovered sculptures that had been exposed by erosion.

Back then, in the 1960s, much of the region was still covered by rain forest, and getting to San Lorenzo required a six-hour boat trip up the Coatzacoalcos River and then a further trek on horseback.

Within hours of his arrival, Dr. Breiner found evidence of a magnetic anomaly at the site. Workers began to dig, and seven and a half feet down, they uncovered a remarkable pre-Columbian figure—a squatting man-jaguar that was probably an Olmec rain god.

During two months at the San Lorenzo site, Dr. Breiner was able to identify more than 100 artifacts, including the two colossal heads. One was found lying on its side, buried 18 feet deep. It weighed 10 tons.

The figures seem to have been buried by the Olmecs and many of the finds were mutilated, but archaeologists still do not know why. It is possible that the colossal heads were created to honor rulers or even champion athletes. Burial and mutilation may have been part of a religious ritual, some archaeologists said.

This year, Dr. Breiner returned to Mexico in late April to help archaeologists survey an untouched Olmec site known as Laguna de los Cerros.

The Laguna de los Cerros site extends to the south of the remains of a large pyramid, which now appears to be a 100-foot-high mound. There are other smaller pyramids and structures, many of which the archaeologists say were religious burial tombs, perhaps of Olmec rulers.

Dr. Breiner brought with him a $25,000 cesium magnetometer built by Geometrics, the Silicon Valley company he founded in 1969.

Mounted on aircraft, magnetometers like this one are able to detect magnetic variations as minute as one part in 100 million. They have been used to find the world's largest reserves of oil, uranium and other minerals.

Dr. Breiner's magnetometer was a portable version, attached to an eight-foot-long pole and connected to a data recorder with a display screen. Carrying the pole, Dr. Breiner systematically marched through dense tick- and snake-infested brush, building a record of the magnetic variations at each point in Laguna de los Cerros. An assistant who wielded a machete walked in front of him, allowing the survey to be made in an array of straight lines.

Later, in his hotel room in the nearby town of Acayucan, Dr. Breiner uploaded the data into his portable computer, building a map of magnetic force lines flowing through the alluvial soil that covers the site.

"What Sheldon does with magnetometer technology is really important in the kind of environment that we have here," said Ann Cyphers, a research archaeologist at the Institute of Anthropological Research at the University of Mexico.

There are many areas of the world where rocky terrain or magnetic minerals would frustrate the magnetometers. Archaeologists have therefore begun to rely on other high-technology imaging techniques, like satellite-based radar, to peer into the earth. But Veracruz, with its rich alluvial soil, is an almost perfect environment for prospecting magnetically for the large Olmec stone figures.

"We hope that someday we'll develop a sophisticated, nonintrusive archaeology that will permit us to learn about these civilizations without disturbing them," said George Stuart, a staff archaeologist of the National Geographic Society.

Laguna de los Cerros is also an example of how difficult it is for scientists to force the ancient Mexican civilization to give up its secrets. The

Olmecs were the first people to live permanently at Laguna de los Cerros, but later occupants reused many of the Olmec artifacts, leaving behind a puzzle that was difficult to unravel.

"Every time we find an Olmec site, we also find at least a second occupation," said Dr. Stuart. He described one site that turned out to be a 1,000-year-old museum.

"It was a society that obviously had its own heirlooms from a much older civilization," he said. "It drove the archaeologists crazy."

Standing atop the central pyramid looking south over what was once a broad plaza of the Laguna site, Dr. Breiner contemplated another riddle. Many years ago at the San Lorenzo site, Yale archaeologist Michael Coe found a thin, polished piece of magnetite that he thought might have served as a compass.

If true, it would be a striking discovery, proving that the Olmecs had discovered the compass perhaps more than 1,000 years before the Chinese.

Dr. Breiner said that the layouts of the Olmec religious sites may offer additional evidence supporting that theory. Although the sites were laid out in a general north-south fashion, the lines diverged suspiciously from true north, he said.

This may have been because the magnetic north pole wanders slightly over thousands of years. Dr. Breiner hoped that by precisely measuring the orientation of each site, he could prove that each site was laid out pointing toward magnetic north as it existed 2,000 to 3,000 years ago.

Dr. Breiner has been pursuing his romance with magnetism for more than 40 years. While working on his master's degree at Stanford in 1960, he approached Varian Associates, an electronics firm in Palo Alto, California, to borrow magnetometers for an experiment he was conducting.

Varian executives made a counteroffer: that the young researcher come to work for their company to help develop new magnetometer applications. He joined Varian and, during the next nine years, came up with dozens of new applications, including the idea of mounting magnets on ski boots to help facilitate the rescue of avalanche victims.

In 1963, he got a call from a scientist at Sandia National Laboratories.

"Let's say we lost something," his caller said mysteriously. "Let's say we lost something at sea."

What kind of metals can your magnetometers detect, the caller asked. It turned out that the Pentagon was frantically trying to find a hydrogen bomb that had accidentally fallen from a B-52 off the coast of Spain.

—JOHN MARKOFF, May 1998

6

# UNDERWATER ARCHAEOLOGY

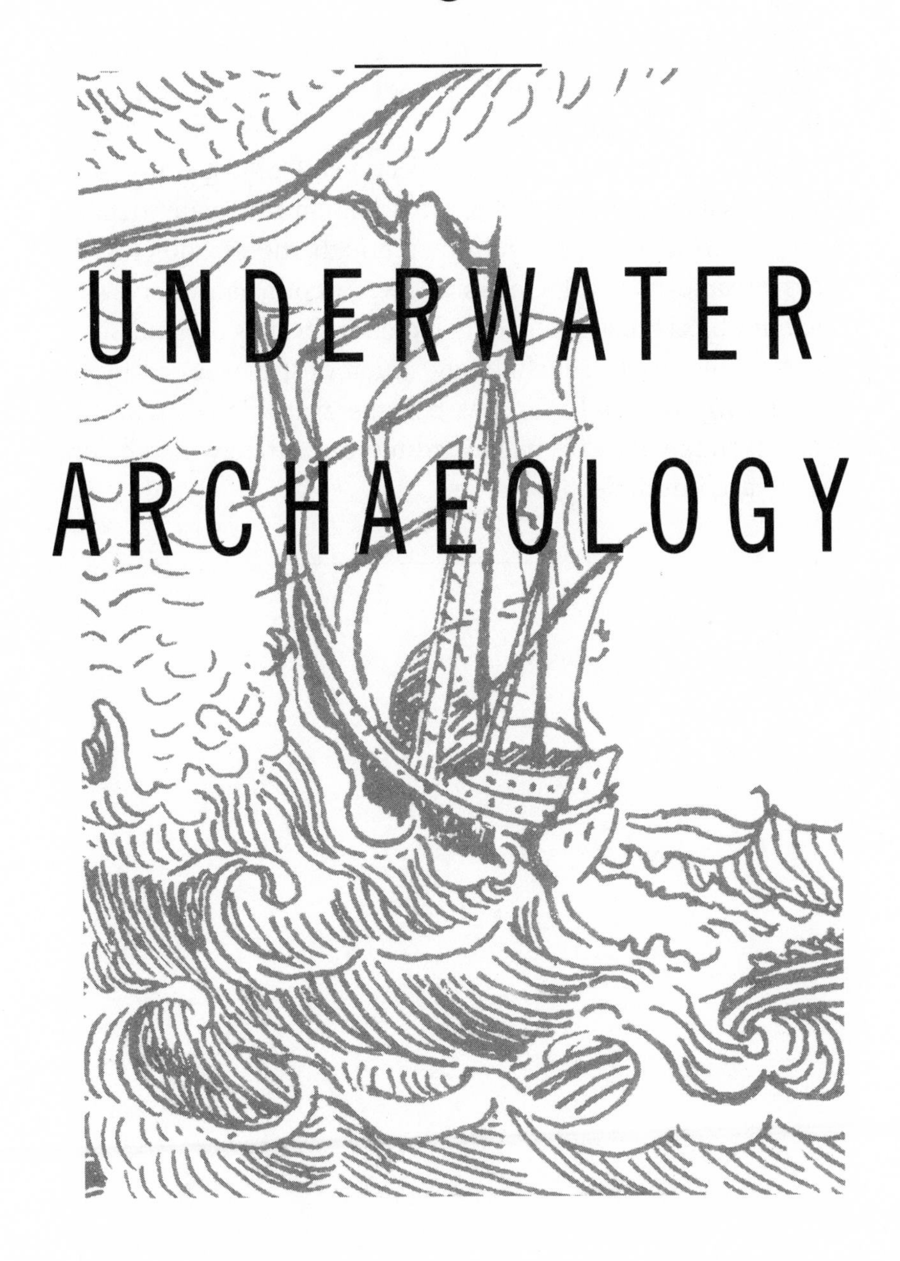

A special branch of archaeology is now coming into its own, the excavation of underwater wrecks, especially from the deep sea.

Ships yield up a unique collection of objects, all dating from the same time and often providing an unusually detailed glimpse into a particular culture. Though the sea quickly destroys organic material, unless at great depth, it preserves the precious objects that are quickly looted from sites on land.

New underwater technology, much of it developed by the navy during the Cold War, has become available to archaeologists, enabling wrecks to be located more accurately than before, and salvaged from deeper waters.

# Secret Sub to Scan Sea Floor for Roman Wrecks

THE MEDITERRANEAN WAS A LAKE to imperial Rome. Hundreds of ships regularly sailed from its seaport town, Ostia, to the far reaches of the empire, ferrying troops and gathering in the oil, wine, slaves, spices, grains and other resources that fed the city. The pace of transport was so great that inevitably, many ships went down in sudden storms, sometimes in deep water where they and their cargoes lay for many centuries, protected from waves and fishing trawls and weekend divers.

These time capsules have caught the attention of Dr. Robert D. Ballard, the marine geologist who in 1985 found the hulk of the *Titanic* under more than two miles of water in the Atlantic. On Mediterranean expeditions in the late 1980s, he discovered a Roman wreck and with the aid of a team of archaeologists recovered some of it, the deepest ancient shipwreck ever to come to light.

Intrigued by that find and the promise of greater ones, Dr. Ballard is preparing to survey and exhume an appreciable part of the watery graveyard that lies along the main trade route between Carthage and Rome, hoping to lay bare some of the secrets of ancient commerce in the Mediterranean.

The tool at his disposal is nothing short of extraordinary—the NR-1, a nuclear-powered, deep-diving submarine that the United States Navy, as part of the peace dividend made possible by the end of the Cold War, is sharing with an increasingly wide circle of scientists. Stranger than anything ever dreamed up by Jules Verne, the once-secret craft has wheels that let it roll across the sea floor as well as windows, lights, sensors, cameras and powerful manipulators for picking up lost objects.

This summer, Dr. Ballard and his colleagues are to glide across the belly of the Mediterranean in the NR-1 to hunt for clues to a bygone world, staying down days at a time, working around the clock in shifts.

"It's perfect for what I want to do," Dr. Ballard said during a recent tour of the submarine at its base in Groton, Connecticut. "This is an unbelievable opportunity."

Archaeologists tend to echo that assessment, saying the unique powers of the submarine promise to shed new light on a significant aspect of mankind's past.

"What's important about this is that, by virtue of the technology, he can get to deeper spots than anybody else," said Dr. John H. Humphrey, editor of the *Journal of Roman Archaeology,* based in Ann Arbor, Michigan. "Previous knowledge of trading routes was based on shallow-water work. This is going to change the whole map."

The Mediterranean voyage marks not only a new venture for Dr. Ballard but a career change as he prepares to move from the Woods Hole Oceanographic Institution on Cape Cod to Mystic, Connecticut, where he is setting up the Institute for Exploration at the Mystic aquarium. The study of ancient trade routes marks the institute's inaugural venture.

Dr. Ballard is known for discovering the wreck of the *Titanic,* the luxury liner, as well as that of the Nazi battleship *Bismarck,* which was sunk in 1941 in the Atlantic and now rests nearly three miles down, a mass of deteriorating guns and faded swastikas.

He first became intrigued with the ancient Mediterranean when, after much effort and frustration, expeditions he led in 1988 and 1989 with deep-diving robots were able to locate heaps of artifacts on the seabed, including many amphoras—the large, all-purpose jugs made of clay that were widely used in antiquity to transport goods.

Even more important was a discovery he made some 60 miles north of Tunis and 20 miles north of a shallow area known as the Skerki Bank. There, a half mile down, Dr. Ballard found a complete Roman ship, its location suggesting it was sailing from Carthage to Rome.

Working with Dr. Anna Marguerite McCann, a marine archaeologist and trustee of the Archaeological Institute of America, he used his *Jason* remotely controlled vehicle to recover 48 of the ship's artifacts, including 10 amphoras, a pottery lamp emblazoned with a running animal, a piece of cedar deck planking, iron anchors, a grindstone, a cooking pot and a copper coin from the reign of Constantius II (A.D. 355 to 361), helping date the wreck to the second half of the fourth century. Most of the artifacts were from North Africa, Italy and the eastern Mediterranean, suggesting a wide swath of travels and trade.

"Ships from this period are the least well known," Dr. McCann wrote in a book-length supplement to the *Journal of Roman Archaeology* devoted to the wreck. "Our late-Roman ship sailing from Carthage is thus a welcome addition to the growing history of ancient seafaring."

Tantalizing the explorers, the seabed around the shipwreck was strewn with evidence of other hulks and a variety of artifacts, suggesting that many thousands of items lay buried in the bottom mud. Dr. Ballard and his team were able to retrieve 17 amphoras from the wider region, the earliest one from the fourth century B.C.,during the early Roman Republic, and the latest one dating to somewhere between the ninth and twelfth centuries and probably Islamic in origin.

The richness of the field hinted at the existence of a previously unexplored trade route over the open sea between Carthage and Rome, experts later said. Apparently, ancient mariners often ventured out of sight of land, contrary to the impression left by decades of work on wrecks in shallow coastal waters.

"They were ambitious," Dr. Humphrey of the *Journal of Roman Archaeology* said in an interview. "The craft they were building were meant to withstand storms on the deep sea."

No less ambitious than the mariners of antiquity is Dr. Ballard, who has talked the navy into letting him use the NR-1 to map the trade route in detail. The submarine has 27 external lights for illuminating the abyss and powerful sonars that sweep outward from its sides, revealing distant objects hidden in the inky darkness.

The world's smallest and deepest-diving atom-powered submarine, the NR-1, or Nuclear Research-1, was launched in 1969 and performed a host of shadowy missions during the Cold War, only a few of which the navy will talk about. In 1976, it helped salvage a sunken navy F-14 fighter and its Phoenix air-to-air missile, the workings of which at the time were highly secret.

Like something out of a spy novel, the sub during the recent tour was moored by the Thames River in Connecticut, low in the water, ominously dark and wet with rain and fitting the somber mood of the gray day. Its interior was warm and well lit but cramped considering the sub measures 146 feet from bow to stern. Somehow, it can shelter and feed 11 crew members and two scientists in an area that feels like a small bus, though a high-tech one packed with shiny pipes and electronic wizardry.

"The habitable area is pretty dinky for a nuclear submarine," conceded Lieutenant Commander David A. Olivier of the navy, the sub's commanding officer. "But it's really pretty large for a deep submersible."

Two chairs in the sub's control area faced a blur of switches and glowing monitors, including ones for viewing the sea floor from the sub's low-light cameras. As if by magic, the floor behind the control center lifted up to reveal a hidden well in which observers lying on their stomachs could peer out of three small portholes into an area where a nine-foot mechanical arm could work the sea floor beneath the sub.

Standing amidships in a sweater and turtleneck, Dr. Ballard said he hoped the expedition would achieve "a beautiful map that would take archaeologists a lifetime to take advantage of." His aim, he said, was to survey the wider region rather than dig into it this time around.

On a personal note, he added that he hoped the trade route proved to be very rich and that he might spend the rest of his professional days helping to reveal man's past. "It would be nice," he said, "to find a place that I could dedicate the remainder of my career, working with archaeologists and preservationists and conservationists and doing some really significant work."

Based on the previous expeditions, he estimated the total area of known wreckage at more than 20 square miles.

"It's massive," he said during a briefing back at the sub base, adding that the site had been protected from the ravages of deep-sea fishing trawlers not only by its depth but probably by the nearby presence of the Skerki Bank, which is a notoriously treacherous shoal.

"This has been a trade route for millennia," he said, indicating its path on a Mediterranean map. "It was critical. From Europe to Africa, this was the shortcut."

Deep old wrecks, he said, tend to be better preserved than shallow ones, which are often damaged by waves and other disturbances. By contrast, "deep shipwrecks settle where there is no light, reduced biology and freezing cold," he said. "That inhibits many processes of decay."

Amazingly, he said, the wood of deep shipwrecks was often preserved if it settled into mud and sediment, safe from certain kinds of marine worms and bacteria. "It turns out that, for some reason, keel-shaped hulls will settle so the waterline becomes the mud line," he said. "Most of the ship is under the mud. Only the top is attacked."

Clearly taken with Dr. Ballard's briefing were some of the people who are helping to finance this summer's expedition and his new career. There were several officers of the J. M. Kaplan Fund, a New York City–based foundation whose Exploration and Technologies Program is supporting Dr. Ballard and other efforts of modern archaeology.

Also present was Hugh P. Connell, president of the Mystic Marinelife Aquarium, an arm of the Sea Research Foundation. Capitalizing on growing public interest in the underwater world, the aquarium is in the midst of a $45 million renovation and expansion, with up to $15 million going to Dr. Ballard's Institute for Exploration. It will not only do science but display it to the aquarium's patrons, possibly with live television links to distant research sites.

"No other aquarium has done much with the deep, deep sea because we know damn little about it," Mr. Connell said after the briefing.

Dr. McCann said the new work, if pursued vigorously over the years, promised to open a new chapter on the ancient world. "It's a very valuable thing to do," she said in a telephone interview. "It gives us a new angle on trade over a period of more than a millennium. Anything we can find is tremendously interesting.

"People have focused on the glamour of the ancient shipwrecks, on finding bronze statues and that kind of thing," she added. "But you can see patterns of trade without lifting a thing if you know the ancient materials. There a lot's down there to learn."

—WILLIAM J. BROAD, February 1995

# Roman Ships Found Off Sicily, New Sites Broaden Study

EXPLORING OFF THE NORTHWEST coast of Sicily with a once-secret nuclear submarine, oceanographers and archaeologists have discovered the largest concentration of ancient shipwrecks ever found in the deep sea, including one ship that may have carried a prefabricated temple.

The findings take archaeology deeper than ever before, promising a new era of discoveries in maritime history.

A research team led by Dr. Robert D. Ballard, whose previous finds include the *Titanic* and the German battleship *Bismarck,* announced the discovery of eight sailing ships lying 2,500 feet beneath the Mediterranean.

With the United States Navy's NR-1 nuclear submarine, the explorers were able to reach 3,000 feet and search the bottom for weeks at a time, using long-range sonar to detect shipwrecks at great distances. Then, with the remotely controlled vehicle *Jason,* which can descend 20,000 feet and use grappling arms to collect artifacts, the team inspected the wrecks up close and retrieved 115 items from the oldest ships.

Because the artifacts were found in international waters, they presumably belong to the salvagers under maritime law. Their historical value is inestimable, their monetary value unestimated.

Five ships were from Roman times, presumably lost in storms while plying the busy trade routes from Rome to North Africa. The oldest, a 100-foot-long vessel dating from about 100 B.C., is one of the earliest Roman wrecks ever discovered. Her holds were filled with amphoras, the clay shipping containers of the ancient world. Another Roman ship, probably from the first century A.D., carried cut stones, apparently ready for assembly into a temple.

Also in the wreckage, which is spread over 20 square miles, were three more modern sailing ships. One was an Islamic ship from the eighteenth or early nineteenth century; the other two were lost in the nineteenth century.

Dr. Ballard, president of the Institute for Exploration, in Mystic, Connecticut, and other team members described the discovery in interviews and at a news conference at the National Geographic Society in Washington.

Recalling the view of the wrecks from the submarine, Dr. Ballard said in an interview, "All of sudden, we realized we had found a graveyard of ships spanning two thousand years."

Dr. Anna Marguerite McCann, the expedition's director of archaeology, was enthusiastic about the immediate discovery, but even more so about its implications for the future. Both she and Dr. Ballard spoke of the research as a demonstration that deep-sea archaeology is emerging as a new branch of science.

When marine archaeology was limited to 200 feet, 95 percent of the sea floor was beyond its reach. Archaeologists and historians often assumed that mariners in antiquity hugged the coast and rarely ventured into the open sea. Their assumption is only now being disproved.

"The high seas were definitely traveled by mariners in ancient times," Dr. Ballard said. "That means the potential for discovery with this new technology is huge. The average depth of the Mediterranean is nine thousand feet, and so there must be a tremendous amount of antiquity preserved in the deep sea."

Not only does the large number of shipwrecks in a relatively small area suggest heavy traffic, but the state of preservation at great depths promises to be especially revealing.

"Deep-sea wrecks are more likely to be intact and their artifacts unbroken," Dr. McCann said. "In shallow waters, they are more likely to get banged up by waves and against reefs, encrusted with coral or buried by sand or looted by treasure hunters."

The wooden decks, rigging and upper hulls of the five Roman ships had been destroyed by wood borers, team members said, but timbers buried in mud were well preserved. The cargoes appeared to be undisturbed. The pilot of the remotely controlled *Jason*, which is owned by the Woods Hole Oceanographic Institution on Cape Cod, was able to maneuver the robot on a detailed survey of each wreck and pick up delicate artifacts, including fine glassware.

Among the items are kitchen and other household wares, fine bronze vessels, two heavy lead anchor pieces and at least eight different types of amphoras intended for storing wine, olive oil, fish sauce and preserved fruit, Dr. McCann said.

One of the most intriguing items is the heavy cargo of high-quality marble or granite building stones still in the hold of a Roman ship from the first century. The artifacts include monolithic columns and large cut blocks of stone.

"Some of the stones appeared to have notches where they were to be fitted together," Dr. Ballard said. "Perhaps this was a prefab temple being shipped somewhere."

Dr. McCann, an adjunct professor of archaeology at Boston University, said discoveries like these promise to provide a new picture of trade in the ancient Mediterranean. "A new economic history of the Roman world is going to have to be written," she said.

In 1989 Dr. Ballard, working with Dr. McCann, made their first ancient deep-sea discovery, a small fourth-century Roman ship they named Isis. The *Jason* robot was used to recover artifacts that suggested it was a widely traveled trading vessel. The location of the wreck, off Sicily, suggested that the ship was sailing between Rome and Carthage, a powerful North African city in antiquity near present-day Tunis.

For the new explorations, Dr. Ballard's expedition returned to the site, about 80 miles northwest of Trapani, Sicily. The waters there are known for their treacherous currents and dangerous reefs.

In some cases, the submarine picked up debris trails on the bottom, evidence of what Dr. Ballard said was a storm-tossed ship in distress that was jettisoning cargo in an attempt to save itself. Sometimes at the head of a debris trail would be the wreck of the ship. A trail that did not lead to a wreck presumably told the story of a storm's survivor.

These trails reminded Dr. Ballard of the story of St. Paul in Acts 27:38: When his Rome-bound ship ran into a storm, "they lightened the ship" by throwing over the cargo.

Further exploration of the Mediterranean routes is planned and next summer, Dr. Ballard said, he is to investigate trade routes in the Black Sea.

—JOHN NOBLE WILFORD, July 1997

# Saving the Ship That Revolutionized War at Sea

THE FEDERAL GOVERNMENT is taking its first major steps to save what remains of the *Monitor*, one of history's most celebrated warships, now a mass of deteriorating wood and metal, half buried in sand, 230 feet down in treacherous waters off Cape Hatteras, North Carolina.

Experts say the famous ship could fall apart at any time. So the government is proposing to lift about a quarter of it, including the engine, the propeller, the massive iron turret, two nine-foot guns and some of the heavy armor belt that still girds the sunken warship. The artifacts would be treated to try to reverse the corrosive effects of time and seawater, then displayed in a museum.

As schoolchildren know, the *Monitor* was an ironclad gunship whose advanced design changed the course of history. It fought a famous Civil War battle in the waters of Hampton Roads, Virginia, against another ironclad, the Confederate ship *Virginia*, formerly the U.S.S. *Merrimack*. Later, the *Monitor* foundered off North Carolina, but its success marked the end of wooden men-of-war and the beginning of the age of armored battleships.

"Within two days of learning the news from Hampton Roads, the Royal Navy, the world's preeminent naval force, canceled the construction of all further wooden warships," said James Tertius deKay, a naval historian, in *Monitor*, a popular history of the warship published by Walker & Company.

The National Oceanic and Atmospheric Administration, which guards the venerated wreck in a marine sanctuary off Cape Hatteras, presented the proposed recovery plan to Congress. Officially, the options include simply trying to preserve the *Monitor* in place as well as hauling it up completely, which would cost more than $50 million.

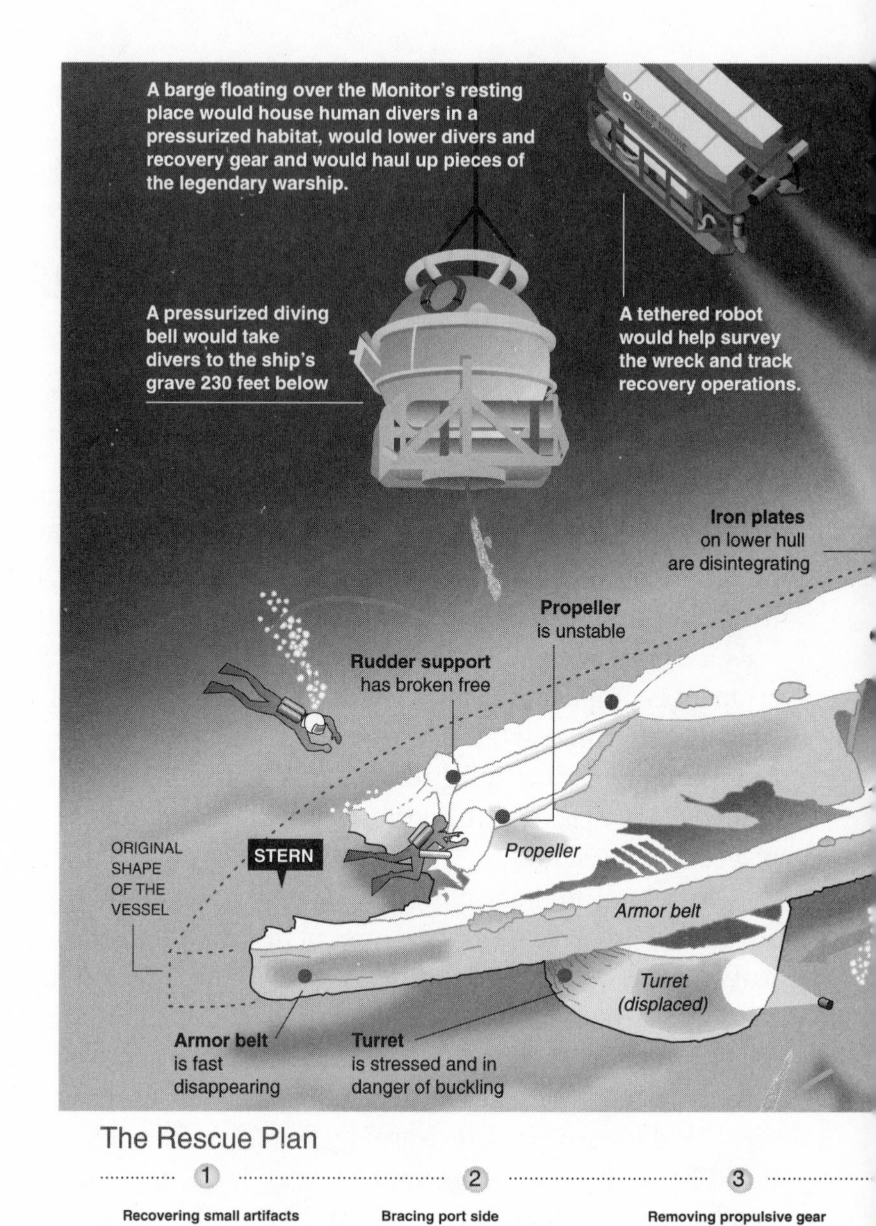


## The Rescue Plan

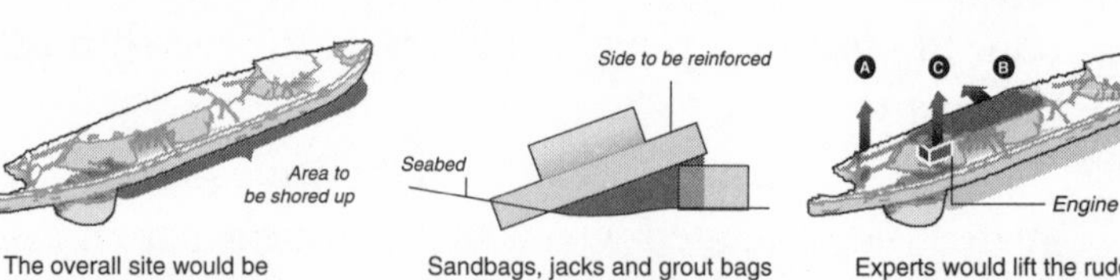


The overall site would be surveyed and all exposed artifacts around the hull removed, especially from areas to be buried.

Sandbags, jacks and grout bags would be used to shore up the ship's port side, which is suspended up to 10 feet above the seabed by the huge iron turret.

Experts would lift the rudder support and two-ton propeller (a), cut open lower hull (b) and remove the steam engine (c).

## ising the Monitor

**er 135 years of undersea decay, and 24 years of expert study, Union warship *Monitor* is in need of rescue. A new plan calls important parts of the ironclad to be raised for conservation l museum display.**

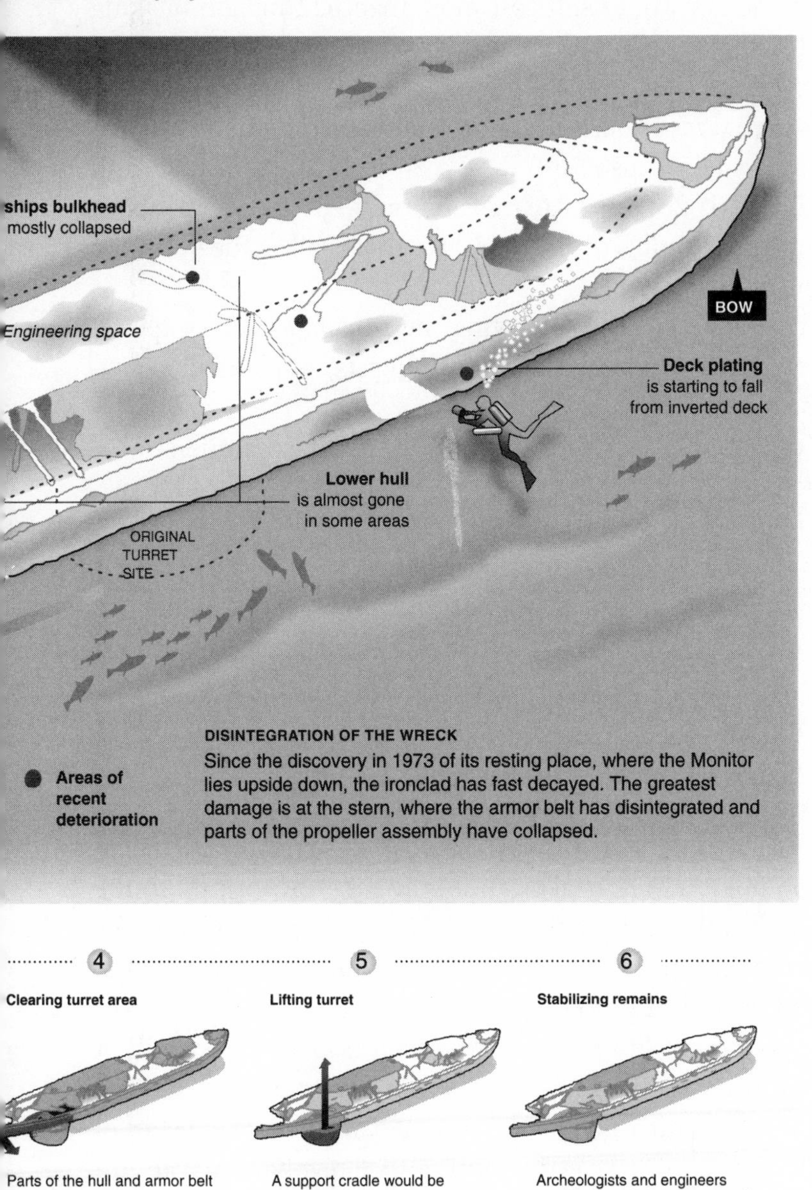


**4 Clearing turret area**

Parts of the hull and armor belt would be removed to give access to the turret.

**5 Lifting turret**

A support cradle would be worked beneath the huge turret and then it would be hauled up.

**6 Stabilizing remains**

Archeologists and engineers would survey the wreck to see if further stabilization is necessary.

The New York Times

Trying to pinch pennies, the agency is proposing a compromise that would save the *Monitor's* turret and some of the ship's stern, which are wasting away after more than a century of briny and human abuse. Though limited in scope, the proposed action would still cost more than $22 million, making the venture a hard sell financially and politically. Experts say that nothing so ambitious has ever before been tried by marine archaeologists; no scholarly recovery effort has been done in such deep water and been so extensive.

A similar effort in 1959 raised the *Vasa,* a Swedish warship that sank in Stockholm harbor in 1628. But it lay in water only 100 feet deep, about the limit for recreational scuba divers and less than half as deep as the *Monitor's* resting place.

Despite the cost and difficulty of saving the ironclad, many experts, including naval historians and marine archaeologists, applaud the proposed recovery as almost obligatory.

"It's time to do something," said Richard W. Lawrence, undersea archaeologist for the state of North Carolina. "We've studied options for two decades. It's time to go ahead and take action."

Mr. deKay said a rescue mission was overdue.

"It is a vastly important American artifact," he said in an interview. "To my mind, it's on the level with the Wright brothers' airplane. It was certainly the most important naval technological development of the nineteenth century worldwide."

John D. Broadwater, manager of the *Monitor* sanctuary, said the legendary ship was fading fast.

"Every time we send down a diver, another deck plate has fallen away," he said in an interview. "It's taken us a long time to work up to the word crisis. You don't want to do the Chicken Little thing. But I think we honestly have a crisis on our hands. Every season counts. We have to assume that any season could be our last."

The *Monitor* was a marvel of technology when first sent to sea. Still, it was but one of a growing number of ships clad in armor plating, though clearly the best of its kind and arguably the most advanced fighting ship of its day. Among its innovations were the world's first revolving naval gun turret and one of the first propellers. Low in the water, the *Monitor* also had one of the first flush toilets on a ship and was apparently the first to have all crew members quartered below the waterline.

Its steam engines turned the screw, a ventilation system and the turret, which was 22 feet wide, weighed 120 tons and was made of iron plate eight inches thick. The guns were two of the most powerful ever mounted on a ship and fired solid iron balls weighing 180 pounds.

The U.S.S. *Monitor* was built in the Greenpoint section of Brooklyn and launched into the East River on January 30, 1862; it measured 172 feet long. An emergency project to help counter Southern shipping, the ship was ridiculed by Union skeptics as an iron coffin and by Confederate seamen as a "Yankee cheese box on a raft."

That judgment changed on March 9, 1862. In the waters of Hampton Roads, Virginia, where the James River meets the Chesapeake Bay, the *Monitor,* with 58 crewmen on board, fought a four-hour duel with the Confederate ship *Virginia,* a steam frigate covered with iron plates.

Neither ship sank the other. But the *Monitor* decisively stopped a bloody rampage against Union ships that the *Virginia* had started the day before.

It was history's first battle between ironclads and the most famous naval engagement of the Civil War.

The draw was an important victory for the North because it prevented the *Virginia* from shattering the Union blockade and helped keep France and Britain out of the war. The South hoped to sell its cotton to Europe in exchange for industrial supplies.

After the battle, the major navies of the world quickly copied the *Monitor's* design, building ironclad warships equipped with propellers and revolving gun turrets. Instantly obsolete were tall ships carrying up to 120 guns and anywhere from 500 to 1,200 men.

Late the following December, after repairs and much feting in Washington, the *Monitor* was heading south for blockade duty off Wilmington, North Carolina, when it encountered gales and turbulent seas off Cape Hatteras, an area notorious for treacherous weather.

Foundering, it went down on December 31, 1862, exactly 11 months after its launching. Lost were 16 crewmen and the ship's cat, which a sailor had tried to shelter in the barrel of one of the guns.

For more than a century, the ship lay in an area known as the graveyard of the Atlantic, so numerous are the shipwrecks.

In 1973, scientists hunting 16 miles off Cape Hatteras found the *Monitor's* remains lying upside down, covered in sand and sponges, clams and

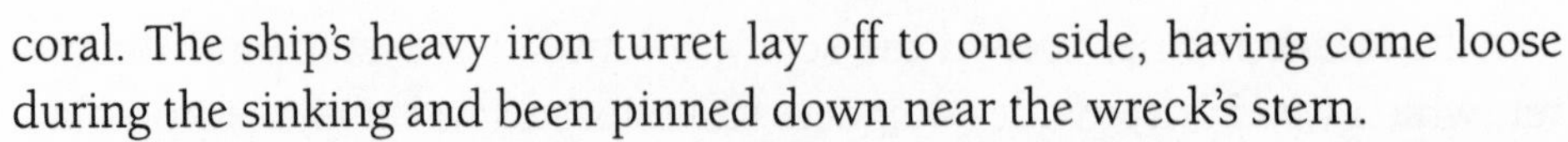


coral. The ship's heavy iron turret lay off to one side, having come loose during the sinking and been pinned down near the wreck's stern.

An analysis of the remains suggested that in World War II, navy ships had mistaken the ironclad for a German submarine and dropped depth charges, blasting apart hull plates and beams.

Overall decay was hastened by the ship's resting place, where the warm Gulf Stream meets icy currents, agitating the waters. The constant churning, abetted by a rich supply of dissolved oxygen, had worked for more than a century to rust some of the ship's massive iron plates.

In 1975, the Secretary of Commerce, who oversees the ocean agency, designated the remains of the *Monitor* as a National Marine Sanctuary, the first of a dozen such protected areas. In 1983, the ship's anchor, weighing 1,300 pounds, was recovered and placed on display in the Mariners' Museum in Newport News, Virginia.

In 1991, a private fishing boat anchored on the wreck. Later analysis suggested that the incident damaged the *Monitor*'s hull. And commercial fishing gear has repeatedly been found tangled in the hulk over the years, signaling the threat of more damage.

In 1995, scientists planned to lift the *Monitor*'s two-ton, four-bladed propeller, which is nine feet wide, but they were foiled by a hurricane.

Over the years, more than 150 small artifacts have been raised up, including wine bottles, condiment and apothecary jars, an ironstone dinner plate, brass oarlocks for the *Monitor*'s boats and the remains of an iron hull plate.

A lengthy report on the ocean agency's new recovery plan, dated October 1997 and given to Congress in November, says that since its discovery, the *Monitor* "has suffered notable deterioration of almost every portion of its hull, with the most extensive damage occurring in the stern."

The rescue bid would involve a flotilla of advanced gear. The salvage project would include undersea robots and a floating barge equipped with cranes, winches and a special habitat for deep divers. Living under pressure and breathing special gases that curb illnesses, divers would descend to the wreck in a diving bell. From there, they would plunge into the cold water to work for extended periods on the celebrated wreck.

In the first phase, the entire shipwreck would be carefully mapped, and smaller items from areas expected to become inaccessible later would be recovered for conservation and display.

Next, the goal would be to shore up the ship's port side, which is suspended up to 10 feet above the seabed by the massive iron turret. The gap is stressing the hull and threatening to break the wreck in two, destroying much of what remains. Divers would fill the gap with such things as sandbags, pumped-in sand, mechanical jacks and bags of grout, which would harden like cement.

Archaeologists and divers would then do the main rescue work, first recovering the propeller and engine, then the turret and its guns. The turret would probably fall apart under its own weight so the plan is to lift it in a support cradle.

Sections of armor belt and hull removed during the recovery process would either be left on the seabed or lifted up to the surface for conservation.

Most artifacts would go to the Mariners' Museum, which already displays the iron anchor and a brass signal lantern as well as *Monitor* plans, photographs, illustrations and memorabilia. Other items might go into traveling exhibitions.

Experts and advocates say the plan is heavy with uncertainties, not the least of which is Mother Nature.

"There are strong currents and sudden storms," said Mr. Broadwater, the sanctuary manager. "It's just not the kind of place that's conducive to long operations. That's a worry."

Another question is whether the recovery can be pulled off even if the weather cooperates. Similar recovery efforts have failed: The Civil War gunboat U.S.S. *Cairo,* sunk in the Mississippi River, was ripped apart in 1965 when cables slung under its hull cut through water-damaged wood.

The biggest challenge is financial. The whole sanctuary program of the ocean agency has a budget this year of $14 million, and the *Monitor* effort, at $22 million, would dwarf that. So the agency is looking for public and private allies to contribute money, gear and enthusiasm.

Officials are upbeat.

"We've been pleasantly surprised by the extent and the intensity of the support," said Stephanie Thornton, head of the sanctuaries program. "There's tremendous interest out there."

One potential partner is the navy, which in theory could lend a lot of assistance. So far, however, it is offering only analytical aid.

Experts say the service may eventually come around to playing a larger role in the effort to rescue the legendary ship that it built.

"The navy is proud of its high-tech history," said Mr. deKay, author of the *Monitor* book.

"I sense that they would do anything to bring her up, to save her in whatever form that can be done."

—WILLIAM J. BROAD, December 1997

# Watery Grave of the Azores to Yield Shipwrecked Riches

THE AZORES IN THE AGE of exploration were an obligatory last stop for ships returning to Europe with the wealth of the Americas and the Orient, full of gold and silver, silks and spices, gems and diamonds, porcelains and fine steels. Over the centuries, thousands of galleons and other ships stopped at the lush volcanic isles for rest and refreshment, bracing for the final push home—often to no avail.

Attacked by pirates, destroyed in battle, ravaged by storms, many hundreds of vessels sank to form a hidden museum off-limits to even the deepest divers.

Until now. The doors are opening for the first time, exciting salvors and archaeologists around the world as they jostle for position in line. By some estimates, the Azores' deep waters hold not only the world's greatest concentration of treasure wrecks but countless warships and artifacts that offer an unusual window on Western history and development.

"It's a turning point," Dr. Margaret Rule, a prominent archaeologist in England, said of the opening, which she is aiding. "The Azores were the crossroads of the Atlantic."

Robert F. Marx, a treasure hunter who has investigated hundreds of shipwrecks around the world, called it an archaeological prize. "I've found so much gold and silver in my life that I'm sick of it," he said from Lisbon, where he is vying for a license to explore in the Azorean waters. "This is different. It's my childhood dream come true. It's the only place I know where there's hundreds and hundreds of intact ships. It's history come alive."

Dr. Francisco J. S. Alves, director of the National Museum of Archaeology in Lisbon, said the stakes were extraordinarily high, calling the Azores "a kind of world sanctuary of underwater culture."

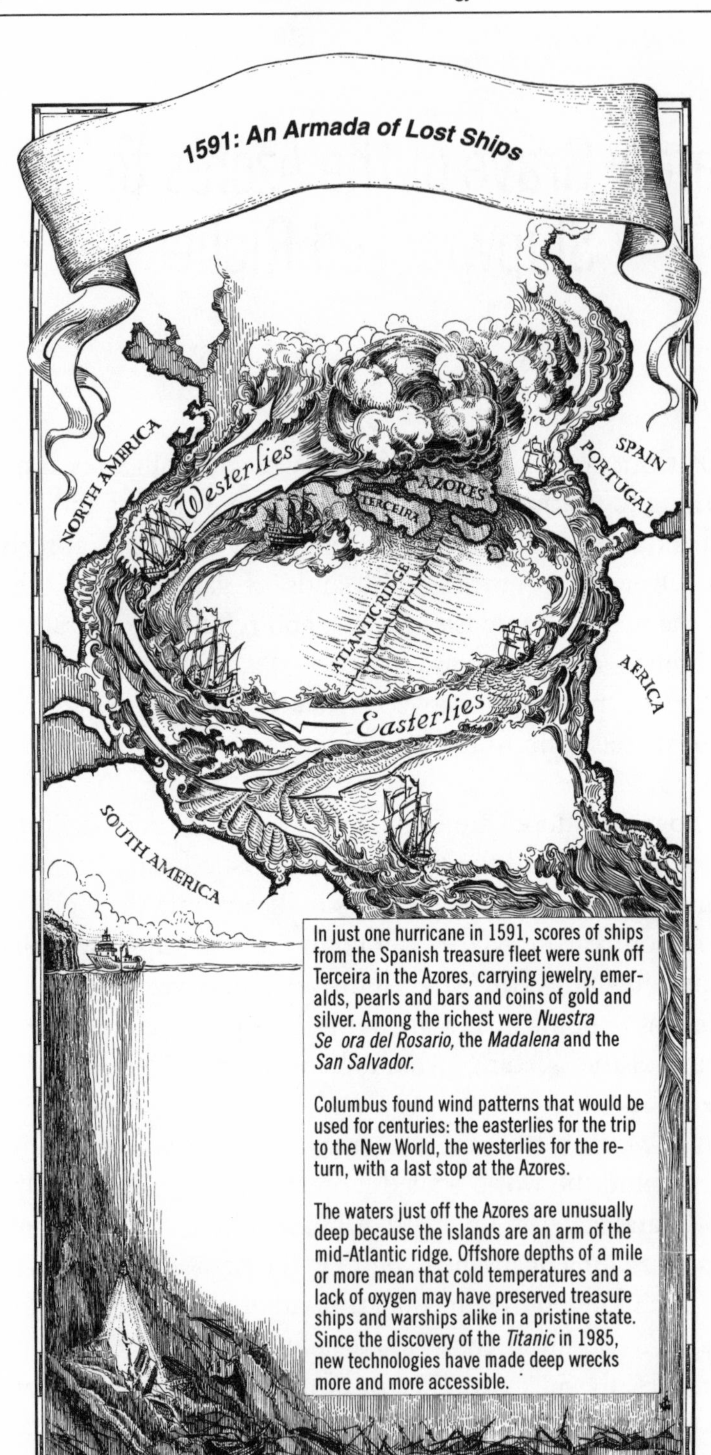


Dimitry Schidlovsky

Protected by unusually deep and icy waters, the shipwrecks of the Azores are now attracting attention partly because technical advances are opening the abyss. Lost ships have long been salvaged in shallow waters, where divers go down 100 feet or more. But deeper ones, like most of those in the Azores, have become accessible only with the wide availability of advanced gear that is beginning to illuminate the sea's inky depths for the first time.

The other factor is Portugal. Proud of its maritime past, happy to aid the Azorean economy, eager to increase revenues from taxes and fees, Lisbon is opening the Azorean depths to commercial exploitation while striving to create a public showpiece that sheds as much light as possible on the nautical heritage of Portugal and the West.

As such, the Azores are a case study in whether an ocean state can foster both private gain and public knowledge, a delicate balancing act that has failed in some countries and invariably upsets partisans.

"The question is whether these riches go to the antiquities markets and private collections or to museums and scholars for scientific study," said Dr. Alves of the National Museum of Archaeology, who has accused the government of wrongly favoring commercial interests over scholarly ones.

But Rui Gomes da Silva, the member of Portugal's parliament who wrote the shipwreck law governing the opening, said the issue was how to strike a middle ground rather than one extreme or another, marshaling as many forces as possible to expedite the opening and enhance the nation's reputation.

"The law is good," he said. "It can make Portugal the center of all the underwater archaeology in the world."

Experts say Lisbon's wreck debate is haunted by the threat of theft, just as in bygone days when pirates looted Spanish and Portuguese treasure fleets.

"The government recognizes that there'll be piracy unless they get control," said Dr. Rule, the British archaeologist. "It's an administrative and practical question of trying to resolve a problem that already exists in Portuguese waters."

Dr. Kevin J. Crisman, an archeologist at the Institute of Nautical Archaeology at Texas A & M University in College Station, who has made inquiries about the Azores, said the potential rewards for scholarship were vast, perhaps including discovery of the sturdy little ships known as caravels, a class that included the *Ni a, Pinta* and *Santa Maria,* but about which little is known.

"It has to be fantastic," Dr. Crisman said of the Azorean wrecks. "For the sixteenth, seventeenth and eighteenth centuries, it's a very rich place in terms of archaeological potential. You even have a good chance of finding a ship from a time before Columbus was in diapers. We have all these questions about the ships of that period."

Twelve hundred miles west of Lisbon, the Azores were stumbled upon by Portuguese mariners early in the fifteenth century and quickly became a vital crossroads because of north Atlantic winds that blow past each other in opposite directions.

Through trial and error, mariners, beginning with Columbus in 1492, discovered that their sails could catch easterly winds in the lower latitudes and ride them across the Atlantic. To return, they simply moved northward to catch the westerlies that blew them back toward Europe—and into the Azores, which are generously spread over the sea and hard to miss.

Treasure fleets returning from the New World were joined by ones coming back from India, China and Japan by way of a long sea route that took them twice around the horn of Africa. These ships returning from the Orient also used the Atlantic's wind system, and the Azores, to speed their homeward journey. Overall, the key port was Angra on the Azorean isle of Terceira.

Perils included hurricanes, fog, pirates and war. In 1591, the English tried to seize a Spanish treasure fleet. They failed, but a hurricane soon sank scores of Spanish ships and tons of treasures. Also lost in that storm was the *Revenge,* a famous English warship that had been captured by the Spanish after an all-night battle and whose brave resistance and loss were later put to verse by Tennyson:

> *"And the little* Revenge *herself went down by the island crags*
> *To be lost evermore in the main."*

The waters of the Azores are unusually deep because the archipelago is a mountainous arm of the mid-Atlantic ridge. Rocky shorelines quickly give way to depths of a mile or more. By some estimates, wooden ships there might be well preserved because of the deep's cold temperatures and lack of oxygen.

For decades modern archaeologists worked only on shallow wrecks, mainly using scuba gear. Supported by public and private funds, Dr. Rule and hundreds of volunteers used such techniques off Portsmouth to salvage the *Mary Rose,* the flagship of Henry VIII that sank in 1545. No treasure was sought or found, but the archaeologists recovered tens of thousands of artifacts that illuminated Tudor life, including longbows, arrows, dice, jars, shoes, clothing, guns and medicines.

The impenetrability of the deep began to change a decade or so ago as new kinds of robots, sensors and submersibles made their debut and gained wide use for undersea exploration. Eventually, the wave included gear cast off by the world's militaries after the Cold War.

The first big breakthrough came in 1985 when the *Titanic* was discovered under 2.5 miles of water. More than 4,000 of its artifacts were subsequently gathered up during three expeditions, with more than 150 of them now on display at the National Maritime Museum in Greenwich, England.

Mr. Marx, the treasure hunter, working with Seahawk Deep Ocean Technology Inc., a recovery company in Tampa, Florida, from 1989 to 1991, salvaged a seventeenth-century Spanish merchant ship a quarter of a mile down off the Dry Tortugas of Florida. Using a deep-diving robot with lights and claws, they picked up not only pearls, gold bars and jewelry but wooden beams, olive jars and ballast stones, recording each item's position for future archaeological analysis. Even some human teeth were gingerly recovered, including the preeruptive molar of a child.

Seeing the opportunity at hand, and urged on by Mr. Marx, Portugal passed a shipwreck law in 1993 that sought to recover and preserve the "undersea cultural patrimony of Portugal."

Under the law, companies are invited to propose search and salvage operations and to prove their expertise and chances of success. If artifacts are judged to be particularly attractive to Portugal, the government has the right to buy them. In general, the reward for the finder is to vary between 30 and 70 percent of the wreck's value, depending on the difficulty of the recovery job and the estimated value of the artifacts.

Exploration licenses cover a maximum area of 100 square miles and recovery licenses cover a zone two miles in radius.

Today, a dozen or so companies and groups are applying to work in Portuguese waters, with Mr. Marx, Dr. Rule and Dr. Alves focusing on Ter-

### A Catalogue of Riches: A Century of Wrecks off Terceira

In all, 867 ships are believed to have been lost in the Azores in the colonial period; of these, 274 were reported lost on or near Terceira Island. Here are some of them:

1504: Caravel loaded with elephant tusks and six chests of gold from Africa.

1517: *Nossa Senhora do Livramento,* with 44 bronze cannons, described then as the richest ship ever to return from the East Indies.

1533: *Nossa Senhora de Piedade,* with 66 bronze cannons and valuable cargo from the East Indies.

1542: Reported total losses up to this date of more than 30 Spanish ships off Terceira alone; individual records were lost in a 1551 fire in Seville.

1542: Portuguese ship returning from East Indies with a treasure of porcelain, gold, silver, precious stones and "other commodities of the Orient."

1544: Flagship of Spanish fleet, laden with treasure, sunk by French pirates.

1549: Spanish ship, the *Santa Barbara,* returning fron Panama with treasures from Peru.

1552: Two Spanish ships, the *Madalena* and the *Santiago,* returning from Panama and Havana.

1555: Portuguese ship, the *Asumpcao,* returning from the Far East, lost in a storm with "a great quantity of treasure aboard."

1560: Two Spanish ships, the *Concepci n* and the *Trinidad,* carrying treasure from Peru.

1586: Two Spanish ships, the *Santiago* and *Nuestra Se ora del Rosario,* with valuable cargoes.

1589: Two Spanish treasure ships returning from the New World, the *San Cristobal* and *Nuestra Se ora de Begonia.*

1590: A Portuguese ship with a valuable cargo from the Far East and a Spanish treasure galleon.

1591: At least 88 Spanish ships lost, most near Terceira in a hurricane; six Portuguese ships returning from the East Indies and English warships also lost.

1593: Six Spanish ships in a treasure convoy lost in a hurricane off Terceira; other ships of the same convoy lost off San Miguel Island.

1608: A Portuguese treasure ship, the flagship of an armada.

1609: A Portuguese treasure ship, the *Sao Salvador.*

ceira. Seahawk is focusing on San Miguel Island, the surrounding depths of which are also thought to harbor many lost ships. Search permits are expected to be issued this season, followed by salvage permits.

New regulations have just been issued that make sure Portuguese museums get a share of the artifacts and that substantially tighten the government's supervisory role, giving it the right to take over salvage sites if it desires and to study all recovered artifacts as long as it wants. The tightening has infuriated Mr. Marx, who says he applauds the museum rule but considers other ones too onerous and is tired of waiting to start work.

But Dr. Rule says the new regulations will aid archaeological integrity. "We're at a rather crucial stage in history to find most everything in the deep and destroy it," Dr. Rule said. "My involvement in the Azores is to find a wreck and do an acceptable job to archaeological standards."

Scholars worry that private companies seeking to turn a quick profit for investors will focus solely on treasures and fail to do detailed recoveries that illuminate a ship's cultural cargo.

But Mr. Marx and like-minded entrepreneurs charge that academic archaeologists are unknowledgeable dreamers and are unable to attract the kind of money needed to pull off deep recoveries successfully. All deep salvages to date, including those of the *Titanic* and the Dry Tortugas wreck, have been accomplished as commercial ventures.

"The academics have never tried deep water," Mr. Marx said. "They haven't found a thing."

Dr. Alves of the national archaeology museum insists that private funds can be raised for such projects, and he is working with a private Lisbon group, Arqueonautica, to do so.

"Universities can work with private donors and people who for humanitarian reasons are interested in giving money for underwater archaeology," he said. "We think that is the way to go, and not the way of treasure hunters."

For Portugal, an added factor giving the wreck issue some urgency is the approach of the world's fair in Lisbon in 1998. The theme of Expo '98 is the world's oceans, and the exhibition will feature a large walk-through aquarium. Planners would love to have a Portuguese wreck on display there, perhaps raised from deep Azorean waters.

Mr. da Silva, the author of the shipwreck law, said government critics were overreacting. He said only a dozen or so of the thousands of wrecks in Portuguese and Azorean waters might be selected for recovery in this century, with the work carefully supervised by a shipwreck commission and the Portuguese navy.

"With good application of the law, and good oversight by the navy, we'll discover a way to be rich in cultural terms," he said. "We don't even know now how they made Portuguese caravels. It's very important for our history."

—WILLIAM J. BROAD, June 1995

# Archaeologists Revise Portrait of Buccaneers as Monsters

PIRATES ARE OFTEN PICTURED as inhuman devils, quick to maim and kill in pursuit of treasure. They fired broadsides into hapless merchant ships, sent captives down the plank and took grim pleasure in torturing victims and even one another. Blackbeard was said to discipline his crew with his bare fists and to have forced one prisoner to eat his own ears.

But scholars in recent years have assailed much of this mythology as misleading or wrong. They find the age of piracy in the late seventeenth and early eighteenth centuries to be peopled by rogues at times less cruel and more egalitarian than previously imagined.

Now newly discovered pirate artifacts are starting to confirm and deepen parts of that revisionist portrait, shedding new light on a lost age.

Spoils appear to have been carefully divided for distribution among crew members, including rare jewelry from the African gold trade. And weapons like primitive hand grenades have been found that appear to have been meant more for intimidating victims and waging psychological warfare than for blasting apart ships.

Many of the discoveries come from the wreck of the *Whydah,* a famous pirate ship sailed by Black Sam Bellamy that sank in 1717 and was found in 1984 off Cape Cod. Moreover, archaeologists announced that they had discovered off the coast of North Carolina the remains of what they strongly believe to be Blackbeard's flagship.

"The finds are opening up a whole new world of real piracy that belies the stories," Philip Masters, the head of the team that found the wreck, said in an interview. "Pirates were nowhere near the monsters they were made out to be."

Archaeologists say the hunt is on for at least two other lost pirate ships, promising to redouble the light already being shed on a class of legendary figures cloaked in centuries of myths and misconceptions. Until archaeologists began excavating the *Whydah* (pronounced WID-da), named after the African "widow bird," or the African port of the same name, there was little evidence available to show how the pirates lived.

"The problem is that pirates moved from ship to ship and often came to a sticky end," Dr. David Cordingly, author of *Under the Black Flag: The Romance and the Reality of Life Among the Pirates* (Random House, 1995), said in an interview. "We do not have Henry Morgan's cutlass or his articles of clothing. There's been nothing, really. It's like a whole race of people who disappeared off the face of the earth."

The 100,000 items recovered from the *Whydah,* Dr. Cordingly said, are "pretty amazing—all the guns, the masses of African gold, the carpenter's tools and ax. It's fantastic."

Such finds, he said, are having "a direct bearing on our understanding of the great age of piracy."

The period between 1650 and 1725 saw an explosion in lawlessness on the seas, a time when literally thousands of men—and a few women—rebelled against the established order and devoted themselves to plundering fleets of trading ships, often heavy in gold and silver, spices and ivory, fine steels—and slaves.

At first regarded as common criminals, the pirates of the golden age began to be viewed more sympathetically in subsequent decades and centuries. They came to be seen as bold villains and romantic heroes, images developed in literary classics like *Treasure Island* and *Peter Pan* and celebrated in movies full of swashbuckling action and deft swordplay. Of late, though, scholars, drawing on old books and documents, have thrown cold water on much of the mythology. Dr. Cordingly, a former staff member of the National Maritime Museum in Greenwich, England, is considered one of the best of the revisionists.

"Real pirates had no time for such ceremonies" as sending victims walking down a plank, he said in *Under the Black Flag.*

Captives were killed at times, "hacked to death and thrown over the side," Dr. Cordingly said. But former prisoners were also known to testify to good treatment, apparently because some pirates wanted to cultivate a reputation for mercy that would encourage surrender rather than a resolution among their victims to resist unto death.

The typical plunder was not chests of doubloons but, he said, "a few bales of silk and cotton, some barrels of tobacco, an anchor cable, some spare sails, the carpenter's tools and half a dozen black slaves."

Scholars have also found intriguing hints that pirates were not exclusively European, as is often pictured. Perhaps as many as 25 to 30 percent of them were black slaves who had escaped from captivity or been freed by pirate gangs to join in the seagoing attacks on organized trade.

In the last decade or so, students of piracy have started to be aided by the technological advances that are opening the seas to divers and robots, leading to a growing number of archaeological finds in waters shallow and deep.

The first discovery centered on Black Sam Bellamy, who in 1716 began to prey on ships in the Caribbean. Contemporary reports say that in 15 months, he had captured more than 50 ships, including the *Whydah,* a new, 100-foot slaver packed with captive blacks and treasure.

In April 1717, Bellamy sailed the prize to Cape Cod to visit his mistress, but a storm came up that sank the *Whydah* just off the coast, killing nearly all on board.

Some two and a half centuries later, Barry Clifford, a native of Cape Cod, searched for the wreck from a boat by towing an electronic device meant to detect large concentrations of metal. Finding the broken hulk in shallow water, he hauled up cannons, shot, pewter tableware, navigational instruments, a bronze ship's bell (inscribed "*Whydah* Gally—1716"), silver and gold coins, gold bars and gold dust.

In addition, the wreck has yielded nearly 400 pieces of worked gold, African jewelry from Akan peoples living in what is today Ghana and Ivory Coast. Scholars say these jewelry pieces are some of the few authenticated relics from more than five centuries of European trading in African gold.

Kenneth J. Kinkor, research director at the Expedition *Whydah,* a museum in Provincetown, Massachusetts, devoted to the wreck's preservation, said the delicate African jewelry had been "hacked apart so they could be shared equally among the men." That is a disaster for scholars of African craftsmanship, he said, but it is also a demonstration of democratic ideals.

"These crews had established floating democratic commonwealths that were at war with the whole world," Mr. Kinkor said in an interview. The totalitarian image of pirate captains is often wrong, he added, except perhaps in the case of Blackbeard, who "had an unusual level of authority over his men."

Scholars have found standard articles of agreement for pirate gangs in which the captain would get two shares of the loot, the quartermaster one and a half shares and the regular seaman one share, with extra booty paid out to men who had suffered serious injuries in battle.

The *Whydah*'s bell, Mr. Kinkor said, also suggested a distribution of power. In the era of tall ships, the bell was the soul of shipboard life, marking time and setting rhythms. The *Whydah*'s was found not toward the stern, a traditional site near the captain's cabin, but "forward, where the crews were."

The *Whydah*'s armament also sheds light on pirate tactics. The team has recovered 27 cannons but relatively few cannonballs, suggesting little interest in broadsides meant to splinter wood and sink ships. Instead, some cannons were found packed with sacks of musket balls—a good way to wreak havoc among opposing crew members but leave the ship relatively unscathed.

The *Whydah* has also yielded dozens of hand grenades, hollow spheres packed with gunpowder that were apparently thrown onto the deck of ships as weapons of terror, meant to sow panic more than slay men.

"These guys paid an awful lot of attention to weaponry," Mr. Kinkor said, adding that the pirates tried whenever possible to use bluff and bluster to try to get a victim to surrender without a fight.

Mythology has pirate boarding parties laden with pistols in their belts. But the *Whydah*'s excavators found no evidence of that. Instead, pistols were found tied to ribbons.

"They'd take a length of silk ribbon, attach a pistol to each end and hang the ribbon over the back of the neck so that the pistols would be hanging at waist level for easy access," Mr. Kinkor said. "It's a simple thing, but it gives us an insight into how they lived."

Mr. Clifford, the wreck's discoverer, said in an interview that only 10 to 15 percent of the wreck had been excavated so far. New sonar technologies that fire sound waves deep into the sandy bottom are to aid the hunt this summer.

"We're hoping to find the mother lode," Mr. Clifford said. The ship is thought to have carried four and a half to five tons of silver and gold, he added.

This summer off Beaufort, North Carolina, experts from a private company and the state of North Carolina are to study what they strongly believe to be the shattered hulk of Blackbeard's flagship, *Queen Anne's Revenge,* which sank in June 1718 and was discovered by a team towing an underwater metal detector.

Blackbeard was one of history's most feared pirates, conducting a reign of terror along the Atlantic seaboard from 1716 to 1718. The study of his flagship is expected to yield many insights, including perhaps confirmation that the pirates in one famous action sought to get medicine to treat the syphilis that tormented them.

For a week in May 1718, Blackbeard blockaded the harbor at Charleston, South Carolina, with *Queen Anne's Revenge* in the lead of his pirate fleet. He took a number of prizes. And, by some accounts, he took hostages who were ransomed only when the governor turned over a chest full of medicines, including the syphilis treatment, which in those days would have been an unguent of mercury.

While moving north up the coast after the blockade, the flagship became grounded in June

> In May 1718 Blackbeard blockaded the harbor at Charleston, South Carolina, for a week, with *Queen Anne's Revenge* in the lead. In June, while coming up the coast after the action, the flagship ran aground on a sandbar as it tried to enter Beaufort Inlet. Eventually it sank.
>
> After a spell of debauchery and feasting on Ocracoke Island southwest of Cape Hatteras, Blackbeard was set upon by a force of English troops sent from Virginia, who killed him on November 22, 1718, in a bloody engagement. The pirate was beheaded, and the victors hung his head from the bowsprit of a conquering ship.
>
> Mr. Masters began his hunt for the lost flagship in 1986, and early in 1987, in the rare-book room of the New York Public Library, he found what he characterized as the key to the wreck's location. In an appendix to a 1719 book about a pirate trial, he found details of the sinking of *Queen Anne's Revenge.*
>
> —William J. Broad, March 1997

1718 on a sandbar off Beaufort Inlet. It eventually sank, even while Blackbeard and his men escaped to terrorize new victims on land and sea and hold a famous drunken orgy.

"We're hoping to get the medicine chest," said Mr. Masters, the president of Intersal, a private company based in Boca Raton, Florida, that discovered the wreck. "We think we're going to find a treasure trove of historical information" but little booty, which is thought to have been carried off *Queen Anne's Revenge* before she sank.

Archaeologists say they are searching for a second lost ship of Blackbeard's, the *Adventure,* which is also believed to have sunk off Beaufort in relatively shallow water.

A third Blackbeard hulk also lies somewhere off the coast of North Carolina, said David D. Moore, a nautical archaeologist at the North Carolina Maritime Museum, a state institution at Beaufort. He declined to discuss its name or probable location, citing the possibility of looting by modern pirates. "With the current hoopla over Blackbeard, people will want to go find it," he said. "We need to do it on our own schedule, archaeologically, methodically."

Most books on pirates tend to perpetuate the myths and legends, he said. "Now we're getting an opportunity to look into the past and answer all sorts of questions," he said.

—WILLIAM J. BROAD, March 1997

# Toppling Theories, Scientists Find Six Slits, Not Big Gash, Sank *Titanic*

HUNDREDS OF BOOKS have been written about the *Titanic* and why the opulent liner sank in 1912 on its inaugural voyage, taking some 1,500 lives in the worst maritime disaster of the day. Everyone agrees that an iceberg was the proximate cause. But the nature of the damage that led to the appalling loss of life has stirred debate for 85 years, the issue sustained by a nightmarish sense of disbelief.

How could a ship so costly and so well constructed—the biggest and supposedly safest vessel then afloat, one hailed as unsinkable—turn out to be so extraordinarily otherwise? Why did the *Titanic* go down so fast? Was there no way to avoid the disaster?

A persistent theory is that the iceberg tore open a 300-foot gash in the side of the 900-foot-long luxury liner. But the ship was lost off Newfoundland in waters some two and a half miles deep, and no author or naval detective was able to resolve the mystery.

Even after the liner was found in 1985, expeditions that probed the icy darkness of the deep sea tended to focus on the sheer spectacle of the ghost ship rather than the nature of the wound or wounds inflicted by the iceberg, partly because the bow was mired in mud, hiding the damage.

Now, an international team of scientists and engineers that repeatedly dove to the *Titanic*'s remains in August 1996 is unveiling a surprise answer likely to end the long debate.

Peering through the mud with sound waves, the team found the damage to be astonishingly small—a series of six thin openings across the *Titanic*'s starboard hull. The total area of the damage appears to be about 12 to 13 square feet, or less than the area of two sidewalk squares.

## New Insight into a Great Wreck

Many *Titanic* researchers long believed that a big gash torn by an iceberg was responsible for flooding six of the "watertight" compartments and sending the ship to the bottom. Such a gash was observed by the team that found the wreck in 1985. Now, however, new sonar images of the hull, which is hidden in mud up to 55 feet deep, indicate that six narrow tears let in high-pressure seawater, sinking the ship in a little more than two hours.

### Revised Countdown to Disaster

New undersea studies and computer calculations are helping reveal the detailed steps of the liner's death.

**11:40 P.M. to Midnight**

The *Titanic* strikes an iceberg, which damages plating in six of its 16 watertight compartments. The ship starts taking on water in the bow through openings that average 20 feet below the waterline. By midnight, mail bags are floating.

**Midnight to 12:50 A.M.**

The ship's captain and main architect tour the damaged compartments. The architect warns that the ship has no more than two hours afloat, and that evacuation of passengers should begin immediately. Lifeboat 7 is lowered, filled to less than half of its capacity. The first distress rockets are fired.

**12:50 A.M.**

The forward bulkhead of Boiler Room 5 falls, letting in waters from the ship's fore compartments.

**1:20 to 1:30 A.M.**

With only a few lifeboats left and 1,700 people still on board, panic overcomes many. The wireless is still running on emergency battery power. "We are sinking fast," a desperate message says. The flooding of Boiler Room 4 is well advanced, but there is no flooding yet reported aft of this space. Almost 31,000 tons of water have now entered the hull.

**After 1:30 A.M.**

Water begins to spill into Boiler Room 4 from the top of Boiler Room 5, helping drive the bow down and lifting the stern high in the air. Great strains develop in the ship's midsection, and around 2 A.M., the ship starts to split in two, contrary to reports later received from all the ship's surviving officers, who said it remained intact as it sank. At 2:20 A.M., the great ship disappears beneath the icy Atlantic.

What doomed the ship was the unlucky placement of the six wounds across six watertight holds, the experts say. A different pattern of damage might have avoided the disaster that started late on April 14, 1912, a quiet Sunday evening notable for its clear sky, chilly air and calm sea.

"*Titanic* was a victim that night," William H. Garzke, Jr., a naval architect who aided the analysis, said in an interview. "Everything that could go wrong, did."

Working with computer simulations of the disaster and metallurgic analysis of retrieved fragments of *Titanic* steel, the team also addressed how the ship flooded, broke in two and plunged to the bottom. Finally, the team investigated the likely fate of the rusting hulk in the decades ahead, examining the onslaught of metal-loving microbes.

The group of experts was assembled by the Discovery Channel, which visited the wreck during the monthlong expedition. Mr. Garzke, a member of the Marine Forensics Panel of the Society of Naval Architects and Marine Engineers, a Jersey City group that advised the Discovery Channel on the investigation, was one of the expedition's main experts.

Another was David Livingstone, an official of Harland & Wolff in Belfast, Northern Ireland, the builder of the *Titanic*. It was the first time anyone from the company had descended to the broken hulk.

"The stern is a terrible mess," Mr. Livingstone said over an undersea microphone while exploring the wreck. But the bow, he added, "is still a very beautiful structure."

That remark was heard by a reporter who visited the expedition for about a week.

The opening of the *Titanic* to forensic analysis is part of a global trend in which the end of the Cold War is accelerating deep-sea exploration as former military personnel and technologies enter the civil sector and start to engage in commerce. In this case, the French government's submersible *Nautile* (French for "nautilus," the sea creature that dives into the deep) carried the investigators down to examine the ship's remains.

When the *Titanic* headed out across the Atlantic on April 11, 1912, it had every luxury: a gymnasium, caf s, squash courts, a swimming pool, Turkish baths, a barbershop and three libraries. The first-class lounge was styled after the palace at Versailles. The menu in the first-class dining saloon that fateful night included roast duckling, foie gras and Waldorf pudding.

After hitting the iceberg, the ship went down in a little more than two and a half hours, and the 700 survivors gave conflicting accounts of what happened. Based on eyewitness reports, it was generally believed that hull damage extended from the first through the sixth of the ship's 16 watertight compartments.

The *Titanic* was designed to survive the flooding of three and possibly four compartments, depending on which ones filled up.

At the British inquiry in 1912, Edward Wilding, one of Harland & Wolff's naval architects, proposed that the uneven flooding in the six compartments meant each had suffered unique, uncontinuous damage. Mr. Wilding also proposed that the actual cuts might be relatively small. His testimony was widely ignored. Nearly everyone believed that the only thing that could undo a ship so big and well constructed was a huge gash.

That idea held sway even after the ship's discovery. A Russian expedition to the sunken liner in 1991 studied the ship's hull plate, finding it quite brittle. Dr. Joseph MacInnis, author of *Titanic in a New Light* (Thomasson-Grant, 1992), wrote that repeated strikes to brittle plates perhaps caused them "to disintegrate, one after the other—in effect, opening up the side of the ship."

While leaving many puzzles, the dives to the *Titanic*'s resting place from 1985 to 1995 documented that the ship's bow and stern had come to rest nearly a half mile apart. The ooze between them was found to hold thousands of artifacts spilled from the ship's decks and innards, as well as twisted shreds of hull plating easily retrieved for analysis.

The dives to the wreckage were done by the French state oceanographic group, Ifremer, for Institut Français de Recherche pour L'Exploitation de la Mer. Ifremer in 1985 worked with an American team from the Woods Hole Oceanographic Institution on Cape Cod that originally found the lost liner.

A major aim of the expedition was to use sound waves to image the *Titanic*'s hidden bow. The wreckage, which slammed into the bottom at high speed in 1912, is today said to be buried in up to 55 feet of mud.

Working from the bright yellow 26-foot French submersible, Paul K. Matthias, president of Polaris Imaging Inc. in Narragansett, Rhode Island, imaged the sunken liner with an acoustic device known as a subbottom profiler, working in much the same way that a doctor examines a pregnant woman with ultrasound.

## Faulty Rivets Emerge as Clues to Titanic Disaster

Two wrought-iron rivets from the *Titanic*'s hull were recently hauled up from the depths for scientific analysis and were found to be riddled with unusually high concentrations of slag, making them brittle and prone to fracture.

In a report and interviews, Dr. Timothy Foecke said the slag content of the rivets was more than three times as high as is normally found in modern wrought iron, making it less ductile and more brittle. While it is not clear whether a better grade of rivets would have saved the ship, he said, the developing evidence points in that direction.

It is also unclear whether such high concentrations of slag were typical for the era of the *Titanic*'s construction, from 1909 to 1912. But some historical evidence suggests that the excessive levels might have been abnormal, raising the issue of culpability.

The *Titanic*'s builder, Harland & Wolff, would make no comment on the findings, saying the topic was too old. "We don't have an archivist or anything like that," said Peter Harbinson, a spokesman for the company, in Belfast, Northern Ireland. "We don't have anybody in a position to comment."

First he made images of the port side, establishing an analytic baseline. Then he examined the ship's starboard side, finding a series of six thin openings.

"There's no gash," he said in an interview. "What we're seeing is a series of deformations in the starboard side that start and stop along the hull. They're about ten feet above the bottom of the ship.

"They appear to follow the hull plate," Mr. Matthias added, suggesting that iron rivets along plate seams probably popped open to create splits no wider than a person's hand.

The longest gap, 36 feet from end to end, extends between boiler rooms No. 5 and No. 6, just crossing the watertight bulkhead.

The gaps are small. But in 1912, just after the collision, they would have averaged about 20 feet below the waterline, where high pressures forced seawater through them like jets from firemen's hoses, filling the ship's interior with some 39,000 tons of water just before the sinking.

Mr. Garzke, a senior naval architect in the Washington office of Gibbs & Cox, Inc., a prominent New York firm of naval architects and marine engineers, said the pattern of damage would have been very different if the ship had been moving slower than its estimated speed of 22 knots, a very fast clip.

If going half as fast, the force of the iceberg's impact and the extent of its damage across the hull plates would have both been much less, Mr. Garzke said, adding that fewer compartments would have probably flooded.

"The ship might have just survived," he said.

One old debate is whether the ship broke up at the surface. Some passengers testified to a wrenching split just before the final plunge, while the surviving ship's officers testified that the vessel went down intact.

Gibbs & Cox did a computerized analysis of the issue, based in part on new findings of the metallurgical makeup of *Titanic*'s steel. Like most steels back then, it turned out to be riddled with sulfurous inclusions that sapped its strength.

The computerized study, known as a finite element analysis, suggested that the stresses in the *Titanic* built up sharply as the bow went down and the stern was lifted high in the air. And it found that the strains were great enough to break the gargantuan ship in two, as some passengers reported.

"Once the ship started to fail, be it due to brittle steel, stress, or whatever, you start forming cracks in the hull," said David M. Wood, a Gibbs & Cox engineer who aided the study. Once that happens, he said, "the whole ball of wax changes" as the ship falls apart.

The investigative team tried to discover what happened to the ship on the bottom, finding a huge bend in the port side of the bow suggestive of a very hard impact.

"It looks like it stopped in a hell of a hurry," Mr. Livingstone of Harland & Wolff remarked upon returning from a dive. The *Titanic*'s bow is estimated to have been falling at a speed of 30 to 45 knots when it hit the bottom.

Experts say it is possible that the sudden impact might have enlarged the relatively small iceberg damage. But whether that in fact happened, they add, is still a mystery.

Parts of the wreck that are now structurally intact might collapse sometime in the next century, the thick steel plates melted into rivers of rust, experts say. Thousands of reddish brown stalactites of rust hang down as much as several feet, produced by bacteria and looking very much like icicles.

Dr. D. Roy Cullimore, a microbiologist from the University of Regina in Saskatchewan, Canada, who was part of the investigative team, carefully studied the growths and toured the wreck. He now estimates that the microbes have consumed as much of 20 percent of the *Titanic*'s bow.

"Everything recycles," he said, "absolutely everything."

—WILLIAM J. BROAD, April 1997

# Lost Japanese Sub with Two Tons of Axis Gold Found on Floor of Atlantic

LONGER THAN A FOOTBALL FIELD, the Japanese submarine I-52 was carrying more than two tons of gold and hundreds of tons of other metals and raw materials to the Nazi war machine when it was sighted on June 23, 1944, by Allied forces in the Atlantic. An American bomber swept out of the midnight sky and dropped a torpedo. The pilot, listening to undersea sounds radioed by acoustic buoys, heard an explosion and a metallic groan as the submarine lost air and sank with more than 100 men. Debris found floating the next day included fragments of silk and a Japanese sandal.

For half a century the warship lay hidden in the depths of the Atlantic. No diver, submersible or robot ever uncovered her grave, the exact location of which was a mystery.

Until now. With the help of advanced gear and methods once reserved for the world's militaries, Paul R. Tidwell, a maritime researcher, found the hulk more than three miles beneath the sea. He now plans to recover the gold, valued at more than $25 million, and perhaps eventually to raise the warship as well.

"It's amazing the condition she's in," Mr. Tidwell said in an interview. "There are no rivers of rust like on the *Titanic*," the British luxury liner that sank in 1912 and was discovered in 1985 in icy darkness.

Finding the I-52 is a case study of how the deep oceans, once inaccessible, are being opened up for science and commerce. During the Cold War, deep-ocean technology was a monopoly of the superpowers and their navies. But Mr. Tidwell, using an advanced research ship from Russia and special navy skills from America, put together a team that accomplished something seldom if ever done before by civilians at such depths.

His team located the warship in 17,000 feet of water some 1,200 miles west of the Cape Verde Islands, apparently intact save for a torpedo hole on its starboard side and some minor bow damage. It lies on rocky seabed, sitting upright, lines draped eerily across its deck as if the ghost ship were caught in some kind of preparatory activity.

"We want to disturb the wreck as little as possible," said Mr. Tidwell, a former army infantryman who served two tours of duty in the Vietnam War and won a Purple Heart. "I feel a responsibility to make sure we treat it with respect. Those people were doing their jobs and died bravely, regardless of their nationality."

Jesse D. Taylor, the navy pilot who torpedoed the submarine a half century ago, said the discovery was a surprise. "I never had any idea that this thing could be located," he said in an interview, clearly fascinated. "I just never thought of it."

The I-52 was part of a secretive exchange of materials and technologies between Hitler and Emperor Hirohito. With Allied attacks making surface transport impossible, the Axis powers resorted to submarines sneaking halfway around the globe.

"Japan was desperate for German technology," said Dr. Carl Boyd, a military historian at Old Dominion University in Norfolk, Virginia, who has no tie to the project. "And the Germans were desperate for raw materials."

Three hundred and fifty-seven feet long, bigger than any American submarine of the day, the I-52 was carrying two metric tons of gold, 228 tons of tin, molybdenum and tungsten, 54 tons of raw rubber and three tons of quinine. It also carried 109 men, including 14 experts from such concerns as the Mitsubishi Instrument Company, who were along to study and procure German technology.

The I-52 left Japan in March 1944 with the 4,409 pounds of gold on board, then stopped in Singapore to pick up the other raw materials. In late April it set out again, traveling through the Indian Ocean and around Africa, bound for the seaport of Lorient in Nazi-held France. It traveled the usual way, submerged during the day and surfaced at night, charging batteries.

Unbeknownst to Tokyo and Berlin, the I-52's route and cargo were known to the Allies, who had broken a host of Axis ciphers for secret communications, including German military orders and Japanese naval codes. Plans were laid for its ruin.

On the moonless night of June 23, 1944, under a clear sky, the I-52 rendezvoused with a German submarine in the mid-Atlantic. Food, fuel and two German technicians were taken aboard, as well as a radar detector meant to help the Japanese submarine evade enemy planes as it neared Europe.

But it was too late. Mr. Taylor, then a lieutenant commander flying as part of a naval task force, took off from the aircraft carrier *Bogue* in an Avenger bomber. Near midnight, just after the rendezvous, he picked up the I-52 on his radar. Zeroing in, he dropped flares and two 500-pound bombs and watched in dismay as the submarine desperately sought to dive, kicking up white water and successfully evading the attack.

Laying acoustic buoys over a mile of sea, Commander Taylor and his crew tracked the submarine, the *chu-chu-chu* of its propellers clearly audible. He then swooped out of the sky and dropped his only torpedo. A long silence was followed by a loud explosion.

Far away, both the *Bogue* and the escaping Nazi submarine saw the flares of the distant battle, and both of them noted the position of the blaze above the I-52's grave.

Mr. Tidwell, a maritime researcher who works both for himself and for clients, saw the glimmer of a prize in the emerging field of recovering shipwrecks in 1990 when he stumbled on recently declassified data about the I-52 in the National Archives while tracking wartime gold shipments. The files included decoded German and Japanese radio transmissions about the submarine's cargo.

Excited, he moved to the Washington area with his wife and two children and threw himself into discovering all he could about the lost submarine, eventually obtaining the sighting data of the *Bogue* and the escaped Nazi submarine, as well as an official navy estimate of the I-52's position. Along the way he raised $1 million from investors.

His goal was aided as the world's militaries began to cast off secretive gear after the Cold War, putting robots, sensors, submersibles and novel techniques onto the commercial market, opening the abyss to civilians.

Working through Sound Ocean Systems Inc., a marine contractor and agent in Redmond, Washington, Mr. Tidwell hired a big Russian research ship, the *Yuzhmorgeologiya,* to hunt for the lost submarine with sonars and cameras dangled on long cables. To help run the expedition and refine the Russian sonar data, Mr. Tidwell hired Meridian Sciences Inc. of Columbia, Maryland, a navy contractor skilled in teasing information out of a sonar signal.

Mr. Tidwell also had Meridian do something that in retrospect proved very useful. For the navy, Meridian analyzes secret data on nuclear submarines to remove navigational errors, allowing officers back on land to better understand actual routes. So too, Mr. Tidwell had Meridian do its computational magic on the half-century-old sighting data, in theory giving a better fix on the submarine's grave.

An element of intrigue arose when Mr. Tidwell found that a rival British group, also using a Russian ship, the *Akademik Keldysh,* was planning to go after the I-52 as well. He offered to join forces. But the other group declined and set sail before Mr. Tidwell could. Despite the head start, the British group left the search area empty-handed.

Mr. Tidwell and his team set sail in April. His operations director was Tom Dettweiler, a Meridian staff member who had helped locate the *Titanic.* For two weeks the hunt was fruitless. The *Yuzhmorgeologiya* ran low on fuel and the search was almost abandoned.

Then, on May 2, a sonar reading produced a telltale dark sliver. Another pass with the sonar revealed the sliver to be a sunken submarine lying more than 3.2 miles down in international waters.

"I was under a lot of stress," Mr. Tidwell recalled. "I was responsible for the money. I had been up two and a half days and had cramps in my side I was so stressed."

Everybody was too tired to celebrate.

On May 5, 1995, the team successfully towed a Russian camera sled over the Japanese hulk, photographing it and allowing Mr. Tidwell to make a positive identification. Later, the team also recovered some of the submarine's remains from an adjacent debris field, enabling Mr. Tidwell to make an international claim to its salvage rights.

Remarkably, the submarine was found tens of miles from the navy's estimate of its position and only a half mile from Meridian's calculated spot.

"If we hadn't done the renavigation, they would have come home empty-handed," said David W. Jourdan, president of Meridian and a former navy submariner.

The team is now preparing for follow-up voyages, led by Mr. Tidwell, president of Au Holdings of Centreville, Virginia, his research company. Au is chemical shorthand for the element gold.

Scholars and scientists from American universities and companies are to be invited along on the next voyage to help assess the wreck's condition, and later to help salvage it as well.

"This is going to be hot stuff," said Dr. Boyd of Old Dominion. "It's not the *Titanic* or the *Bismarck,* but in a more subtle way, it's perhaps more interesting because they were pretty well known."

Historians have no photographs of the I-52, one of the biggest of all the Japanese submarines of World War II. By contrast, the *Titanic* was photographed hundreds and perhaps thousands of times, including every stage of its construction.

The I-52's sunken gold consists of 146 bars packed in 49 metal boxes, according to a manifest that was radioed from Tokyo to Berlin and decrypted by its American interceptors. After the recovery, Mr. Tidwell says, the whole I-52 might be lifted by filling the wreck with a special kind of foam, making the vessel light enough to come to the surface.

He says he is working closely with the Japanese authorities and has offered to return any personal effects and possibly the whole submarine, if raised. Mr. Tidwell noted that the Japanese tended to make no claims to lost riches and war booty.

All told, he said, recovery of the I-52's gold might cost $5 million to $8 million, leaving him and his investors a tidy profit. It would also provide the capital for other projects. Without giving any details, he says he has documentary evidence of ships that were lost with five, seven and even 15 tons of gold.

"There was a lot of hardship getting to this point," Mr. Tidwell said of his five-year hunt for the golden submarine. "But now I've reached a plateau of achievement in my life. I was right. And that's a good feeling."

—WILLIAM J. BROAD, July 1995